W9-BVF-542

The Ocean World of Jacques Cousteau

Man Re-enters the Sea

The Ocean World of Jacques Cousteau

Volume 12

Man Re-enters the Sea

THE DANBURY PRESS

The history of man's reentry into the sea goes back thousands of years. Modern equipment and techniques have opened a new frontier to challenge our senses. Truly spectacular sights and adventures await the undersea explorer.

The Danbury Press
A Division of Grolier Enterprises Inc.

Publisher: Robert B. Clarke

Production Supervision: William Frampton

Published by Harry N. Abrams, Inc.

Published exclusively in Canada by
Prentice-Hall of Canada, Ltd.

Revised edition—1975

Copyright © 1973, 1975 by Jacques-Yves Cousteau
All rights reserved

Project Director: Steven Schepp

Managing Editor: Richard C. Murphy

Assistant Managing Editor: Christine Names
Senior Editor: David Schulz
Assistant to the Senior Editor: Robert Schreiber
Editorial Assistant: Joanne Cozzi

Art Director and Designer: Gail Ash

Assistant to the Art Director: Martina Franz
Illustrations Editor: Howard Koslow

Creative Consultant: Milton Charles

Printed in the United States of America

234567899876

LIBRARY OF CONGRESS CATALOGING
   IN PUBLICATION DATA

Cousteau, Jacques Yves.
   Man reenters the sea.

   (His The ocean world of Jacques Cousteau;
v. 12)
   1. Diving, Submarine.   2. Underwater
exploration.   I. Title.
VM981.C68   797.2'3   74-23051
ISBN 0-8109-0586-8

Contents

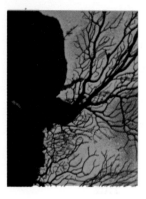

a few minutes, he would need supplies of AIR TO BREATHE (Chapter V). Nature suggested solutions—from the snorkle of some larvae to the diving bell and trapped gas bubbles of the water spider. But tubes could only be used in shallow water; reservoirs of air could not be inhaled by a diver subjected to hydrostatic pressure; and the air from diving bells fouled rapidly. Air compressors provided a solution.

At this point man had learned how to make descents into the ocean and how to extend their depth and duration. But true FREEDOM IN THE SEA (Chapter VI) could not be achieved as long as the diver was still leashed to the surface with air hoses and lifelines. Self-contained breathing systems were the dream of nineteenth-century writers, inventors, and divers. But their inventions were built poorly or not at all, due to the limited technology of the period.

Centuries-long experimentation with better diving methods and apparatus paved the way for modern developments. Man had successfully negotiated the first 200 feet of the ocean world. But to penetrate the sea LONGER AND DEEPER (Chapter VII) some major breakthroughs had to be made, for nature had once again erected a barrier. Compressed air was the limiting factor. New breathing mixtures were required, and the characteristics of artificial atmospheres had to be thoroughly tested before man could risk using them in the sea.

In the slow process of conquering deep waters, man had discovered—often painfully—the laws of decompression. There was little in his own makeup or the nature of breathing gases that he could turn to in the hope of eliminating the chore of decompressing. The discovery of saturation diving, however, opened the door to LIVING UNDERWATER (Chapter VIII). Decompression wasn't eliminated, but was proportionally reduced and postponed. Human sea floor colonies began to sprout, with habitats like Conshelf, Sealab, and Tektite.

Mammals like whales and dolphins have accomplished total reentry into the sea. Man has increased the time he can spend underwater and has even lived in the ocean, yet a return to the surface is inevitable. THE UNDERWATER MAN (Chapter IX), a true creature of the ocean, is still a visionary concept. Work toward this dream is progressing, but it is still theoretical. *Homo aquaticus* remains a being of the future.

To the dark provinces of the sea, a thinking diver is a MESSENGER OF THE SUN.

Introduction: Man's Reentry into the Hydrosphere

Are human beings about to return to the sea as some mammals did a few million years ago to become seals, porpoises, and whales? In the absence of drastic anatomical and physiological mutations, it is very improbable indeed. Our silhouette, our limbs, our lungs, heart, veins, and arteries; our fat and our liver; our kidneys, our skin, our blood—all would have to be modified radically before we could stay submerged for weeks or months at a time, without dying from exposure to cold, losing our skin, or being compelled to go back too often to the surface for air. In spite of the new popularity of skin diving, there is no indication that in the grand scheme of evolution men were programmed to become marine creatures.

However, in his own way, which is an artificial way, man is preparing his reentry into the sea. He makes up for his lack of blubber by developing better diving suits. He struggles both to perfect breathing devices and understand diving physiology to compensate for his inability to feel at home in the water's lap. He has already lived for an entire month in undersea settlements. He is the proud owner and user of dozens of exploration submersibles, and he has gone down in bathyscaphes to depths beyond those ever reached by the sperm whale. In a few decades medical research and surgery may well turn some of us into technological amphibians. Man cannot fly. Man cannot dive very well. Yet he has conquered the air, the moon, and the deepest ocean trench.

The conquest of the hydrosphere is a modern feat—still only imperfectly achieved. But man's futile efforts to penetrate the alien element that nursed his ancestors have been traced to the earliest times. In seaports of the eastern Mediterranean, in the warm waters of the Persian Gulf and the Indian Ocean, in scattered islands in the Pacific, and even in the icy waters of Tierra del Fuego, man was diving before there were scribes to record the event. These primitive divers were at the same time practical and mystical. From the mysterious shallows, they brought back food and treasure, as well as fantastic tales that kept mythology alive—pearls, coral and monster stories, sponges and legends of mermaids. The Sumerian hero Gilgamesh found and lost the fabled seaweed that provided eternal life.

Through empirical knowledge these pioneers perfected naked diving, reaching depths of 150 to 200 feet, in dives lasting two minutes—sometimes longer. Their techniques were passed on from generation to generation. It was only at the end of the nineteenth century that technology and science were able to improve on these methods and develop the equipment necessary to open up a new world for man. Progress in diving was explosive, and coincided with the population and the industrial explosions.

Why have men always been motivated to dive? Was there a subconscious yearning for the element that had produced life, for the mother-sea they came from? Maybe. But the conscious drive was for freedom and adventure. Freedom from their own weight, adventure in exploring a world that can barely be imagined from the surface. The naked apes were compelled to dive, but they were unable to communicate accurately their discoveries and their feelings to their landlocked peers. They were different men, almost sorcerers. Even today, those who have experienced the weightlessness and three-dimensional freedom that can only be experienced in diving and space walking have a secret in common; they have been initiated in the rites of the sea.

Today the motivations are more materialistic. There is a worldwide trend to assess the benefits from research in the same way industry evaluates investments against potential profits. If diving is to progress, it has to demonstrate that it is practical and economical. In the realm of air breathing (surface to 200 feet or slightly more), divers have proven to be irreplaceable: salvage, rescue, coral and pearl collecting, oyster farming in Japan, geological, biological, and behavioral research, speleology, and undersea archaeology have all demonstrated the efficiency of human presence underwater. Deeper diving, however, involves gas mixtures, elaborate habitats and decompression chambers, voice unscramblers, power lines, power tools, and communication devices, and each working hour on the bottom becomes extremely expensive and often dangerous. For this reason, the main customers (offshore oil exploration companies) have developed alternate methods of intervention, using either completely automated tools controlled by closed-circuit television, or capsules lowered from a surface platform and housing men working at atmospheric pressure. Deep divers are in competition with robots.

It would be equally wrong to assert that either deep divers or robots were the best solution in all cases. The same question has been debated endlessly: could direct observation be replaced by instruments in outer space exploration? Experience has proven that both were useful—in inner as well as outer space.

Jacques-Yves Cousteau

9

Chapter I. The Underwater Adventure

Divers are often asked: Why do you do it? Mountain climbers have always faced the same question. And for a long time the response was the same: Because it's there. Mountain climbers still have problems convincing people that there is good reason for hiking up through the clouds, but divers can truly speak of their adventure in the sea.

Compressed-air diving was the boom sport of the 1950s when many enthusiasts took up spearfishing. But as soon as fish started to become scarce, divers then realized that undersea hunting was seriously harming the coastal fish populations, and they turned to other avenues of interest. Resorts in tropical areas began attracting divers with the prospect of observing curious forms of life in

---

"Millions of divers have turned to the sea for freedom and adventure."

---

colorful settings and comfortable waters. Today some coral reefs have been designated underwater parks, where the rule is, of course, "look but don't touch."

Considering that diving started primarily as an economic activity, it is not surprising to find that aqualung divers have turned full cycle and are once again using diving for pragmatic ends. Of course, there have always been military divers, and docks and ships still need underwater repairs, but in the move away from recreational diving there have been some interesting, if logical, twists. There are biology classes that take "field trips" underwater to study animals in their natural habitats. Marine archaeologists use the aqualung as a tool to explore potential excavation sites, and geologists study shore-

line configurations from below the waterline. Medical doctors who treat injured or ill divers have taken to the water to study the stress and strain of diving on the body.

Artists have long been inspired by the sea, and now photographers are actively pursuing their inspiration. Pressurized camera housings, flash attachments, and light had to be developed first. The equipment is constantly improving, and the picture takers have become adept at making adjustments for shooting in a medium that has special optical characteristics. Water's refraction of light and its property of blocking some color lightwaves and letting others pass have been exploited by some photographers who are developing a whole new art form, enhanced by corrected lenses and special lighting.

The proliferation of undersea activities has called attention to the need for some rules and regulations. Divers have adopted the practice—in some cases required by state law—of using a floating flag to warn surface vessels to stay clear of the area. The "diver down" flag is a red square with a white diagonal stripe. The flag is intended to keep boaters at a distance and prevent surfacing divers from being wounded by propellers.

To exchange information and experiences, divers have formed different types of organizations. In some countries, a person must be a member of a recognized club to be permitted to dive. Americans are not that restricted yet. Their clubs are generally formed along interest lines, such as marine archaeology or underwater photography.

In the realm of the underwater world, man is still an alien in a hostile environment, but he is being initiated into three-dimensional freedom.

A jet-propelled octopus spurts by a diver. Many sea creatures avoid man; they have learned that human intruders are usually killers.

Meeting the Residents

Everyone who starts diving must learn a little marine biology, if only to satisfy the curiosity aroused by the myriads of creatures in the sea. An experienced diver knows how to "read" the ocean. Solid bottoms of rock or coral generally support more creatures than sandy or muddy bottoms. Topography itself is important, for there are usually more animals to be sighted in an area with many different levels than in an area with a broad, flat expanse. The rocky crags obviously offer more protection for marine life, although most are well camouflaged and it may take some time to discern them. When swimming slowly and looking closely, one has a better chance to spot many well-camouflaged forms. Still waters may be relatively devoid of creatures, while the presence of a current will bring them out since the current carries planktonic food along with it. If turbulence is too great, however, as on rocks constantly beaten by heavy seas, the population is thin, leaving only the more hardy plants and animals to adapt. Twenty years ago members of an underwater community had no fear of man: they came out to greet divers. They no longer do so, as they have learned that human intruders are usually killers. For the most part they remain in hiding or proceed warily from their shelter. Plants or finned animals may be the obvious sight, but on the underside of rocks crowds of scared creatures may be hiding. In areas where there are few rocks, such as coves and bays, burrowing creatures like some types of anemones, molluscs, and worms tend to concentrate in the sediment.

In warmer waters coral reefs offer one of the most diverse biological communities in the ocean. The reef itself is built by living organisms with many other creatures attached to it or using the open spaces inside it for protection. Another diverse habitat is the kelp bed where many species of fish live among the towering strands and other animals live on the kelp itself. Submarine canyons, those eroded wedges which funnel nearshore sediment out toward the abyss, sometimes provide glimpses of deep-sea fish who follow up the canyon close to the shore.

Some of the more adventurous divers like to confront the larger creatures of the deep and ride them bareback in rodeo fashion. But most divers tend to avoid dangerous animals, and they don't have to be big to be harmful. There are certain species of coral, sea urchins, jellyfish, and cone shells which are capable of imparting stinging venoms.

Stingrays are only likely to use their weapon if one steps on them and sea snakes only bite if they are seriously disturbed by an aggressive diver. There are a number of bright, colorful fish that are venomous. Sharks are dangerous for surface swimmers but rarely attack divers. But at the slightest touch a well-camouflaged stonefish with poisonous spines or a sea wasp jellyfish may inflict extremely serious wounds that have been fatal in a few instances.

The roles are reversed on this diver; a small octopus has decided to investigate him! Undersea venturers should always expect the unexpected.

Getting the Picture

In viewing the spectacular sights of the undersea world, a diver often finds himself at a loss for words to describe what he has seen. He can save himself a thousand words each time he takes a picture.

The man who started all the picture taking was Louis Boutan, a professor of zoology at the Sorbonne. Boutan made his first photographic dives in 1893. He used a heavy helmet diving suit and a fixed-focus camera, which was encased in a copper box and equipped with an external control that allowed exposures lasting from several minutes to half an hour on glass plates coated with collodion emulsions. The pressure inside the watertight copper box was equalized by means of an attached rubber balloon which, when compressed by water pressure, forced air into the casing. There were three plate-glass ports in the housing, two for view finding and the other for the lens.

The major problem to overcome in taking a camera underwater was to provide enough light to penetrate the water, illuminate the

Underwater photographers (above, left, and below) focus in on their subjects. Wide-angle lenses and elaborate lighting systems have allowed the real beauty of the sea in all its glorious color to be captured on film for everyone to enjoy.

subject, and return and expose the film. We consider a water clarity of 100 feet exceptional, but on land such clarity would be dense fog and conditions under which few photographers would work.

As the depth of water increases, the world becomes bluer and bluer because of the selective absorption of the warm colors by seawater. The problem of particles scattering light and blurring the image is also immense. Corrected, wide-angle lenses, as well as elaborate and expensive lighting systems, have been developed to allow the real beauty of the sea to be captured in full color on film, a beauty that no fish has ever seen.

Another problem confronting ocean photographers is the surge. As waves pass overhead, water below moves forward and backward. Any diver wishing to photograph a still subject attached to a rock has the almost impossible task of maintaining position.

If the photographer is concerned with the behavior of an animal, he must contend with its reaction to his presence. Telephoto lenses are of little use underwater because over great distances light will be absorbed or scattered to a considerable extent. Consequently a diver must be very close to his subject; he must make very slow movements and have patience to allow animals to become accustomed to his presence and resume almost normal activities.

For the underwater novice, photography is one of the major opportunities to be of use to science. A picture or film cannot be questioned because of naiveté of the observer. A patient and skilled photographer can record animals or their activities accurately whether or not he knows their Latin names.

Working Underwater

For all the sport diving, spearfishing, and scientific research done by divers, the advances in diving equipment have come largely because of an economic demand. There was work to do, such as salvage operations, harbor and ship repairs, oyster farming, or tunnel building. Even military divers are not used strictly in combat. They have been eventually called for welding and other repair work on the undersides of large military and privately owned vessels.

One such job—trivial at first consideration—is locating intake and exhaust holes on the underside of ships' hulls and plugging them up so work can be done on the inside.

A fanciful salvage attempt is depicted below. Differences in pressure from surface to bottom would never let the diver draw air from above.

Astronauts (above) have even become divers. They learn to function in a weightless environment, attempting to simulate conditions in space.

There are several ways to locate these holes, and the job isn't as easy as it seems. The most primitive method is to have someone on the inside of the ship bang on the hull with a hammer at the spot where the hole is. Then the diver will follow the sound and be guided by its intensity, which can be difficult since water's density conducts sound differently than air. And this technique is virtually useless in a double-hulled ship like a modern aircraft carrier. Another way of locating an opening is to send compressed air through the valve inside so that all the diver has to do is look for the stream of bubbles. But if the hole is located in a flat area of the hull, the bubbles collect on the surface and actually mask the opening. A

*The bright glare from a **welder's torch** (above) is unmistakable. Commercial and military divers are often called upon to do underwater repair work.*

*A **diver** (above) cuts away at the algae-encrusted hull of the Andrea Doria, a 700-foot luxury liner that sank off Nantucket Island in 1956.*

third method is to pour dye through the opening and have the diver follow the color to its source. Since most repairs are done in busy harbors, however, the water is so murky and so much sediment is stirred up that the reduced visibility renders this technique impractical. So a simple task like finding a hole on the bottom of a ship can be an all-day job for a team of skilled divers.

In a recent survey of scientists and engineers involved in oceanographic work, there were only 2 percent who used hard-hat helmets in their work. The basic tools were ships, submersibles, and self-contained breathing apparatus. The many tools available to diving workers are oxyacetylene and oxyhydro-

gen cutting torches, arc welders, bolt-driving and punching guns, and pneumatic drills. Using hand tools underwater is a tricky business, since weightlessness prevents using the same kind of leverage that is effective on land, and any rapid movement, like hammering, is made tiresome by the resistance of water. Divers are often called upon for spectacular tasks, such as recovering space capsules when astronauts return from outer space. On a front closer to home, divers are asked to search harbors or ponds for the body of a drowning victim or to recover lost or missing articles, most often in murky waters where they are practically blind. Yet, under the most dire circumstances, well-trained divers get the job done.

Underwater Speleology

The most dangerous place to dive is inside the earth—into caves filled with fresh water, cold as winter and darker than night. The darkness contrasts with the sunlit waters outside. In such explorations the greatest challenge is finding the way out before the air supply is depleted. In sea caves, underwater flashlights are easy to use because the organisms in suspension in the water give substance to the beam of light. In a freshwater cave, on the contrary, there is no life. The cold water is numbing and strength-sapping. The interior of the cave is jagged, etched by centuries of water scouring the softer stones, sharpening the harder ones. A flashlight is almost useless, for the crystal-clear water seems to eat light unless it falls by chance on the rocks of the wall; but if mud is stirred up, then one is left with no visibility.

In a cavern a diver feels blind and fearful; there are no charts of the twisting, dipping gape within the earth. Orientation is a three-dimensional problem, since a move up may be more dangerous than a descent as the ceiling rises and falls in a pattern of its own. A roof above one's head, making emergency ascents impossible, becomes claustrophobic.

Because of the complexity of the caves and the low-slung siphons leading from one gallery to another, the tools and skills of a mountain climber (especially a good rope unrolled during progression as "Ariadne's thread") are often more useful than are those of a diver to trace a path to the surface.

The sirenlike lure of the cave is there, drawing adventurers to galleries that no man has seen before, or perhaps no man has seen since prehistoric times when his primitive ancestors lived or hunted in the cavern before it was filled with water at the end of some distant ice age many thousands of years ago.

One of the most spectacular finds by a cave diver was at Montespan in France, where Norbert Casteret entered a giant gallery which had been used by hunters thousands of years ago in a semireligious ceremony asking for luck in hunting. Casteret uncovered an effigy of a crouching bear, four feet long and two feet high, with the skull of a bear lodged between the front paws. Apparently the cavemen covered the clay statue with a bearskin and thrust their spears at it for luck. The effigy bore the marks of 20 or 30 spear thrusts. There were also representations of large cats, perhaps lions or tigers, among the 30 clay sculptures in the hall. As Casteret later described the sight, "On all sides carvings of animals, sketches, and mystic signs sprang to our gaze."

Divers (left, below) examine stalactite formations in **submerged caves.** *Cave explorers have made some spectacular geological finds in past years.*

Under the Ice

Winds whip through a frozen landscape on a sunless day. The temperature on Resolute Bay, 600 miles above the Arctic Circle, is 45° F. below zero. A diver prepares to enter water that is 28.5°, colder than ice, but warmer than air. To avoid mechanical ice-ups, he is protecting and warming up his equipment even more than himself.

Dr. Joseph B. MacInnis is leading a team of divers into arctic waters to study man's ability to dive and work in such low temperatures. MacInnis, a physician who has been interested in diving and its effects on the human body, has convinced 14 other men that the experiment is worthwhile, despite the hazards. Accidental exposure can kill a man in ten minutes. Feet and hands risk frostbite. There is a special psychological hazard in diving under ice, for just reaching the surface isn't enough—you can still be trapped under a thick sheet of ice if you miss the diving hole. The procedure used by Dr. MacInnis is to rise to the ice ceiling, then turn upside down and walk on the ice to reach the diving hole, do a flip, and make a headfirst return to the atmosphere.

Half-inch-thick gloves make fine work almost impossible, but they are absolutely essential. And we know that at any time something could go wrong, like a regulator freezing in the closed position. There is as yet little lore about diving under ice, every encounter is a totally new experience. One of the first startling revelations is the abun-

A diver (below) inspects the surface ice covering. Perhaps the diver's greatest worry is missing the opening of the hole and being trapped.

Diving in polar seas (above) *requires special suits to insulate the body. Accidental exposure could kill a man in only ten minutes.*

dance of life in polar seas. There are not many fish, but invertebrates abound. And though there are not very many different species, there are a surprisingly large number of individuals in any one species.

The single most important factor, of course, is keeping the body warm or, more correctly, reducing the amount of body heat lost to the water. One type of diving dress used for this purpose is the Unisuit, a watertight dry suit which includes nylon fur underwear and a quarter-inch neoprene foam outerwear with nylon lining. Both inner and outer suits are one-piece units, complete with feet. Tests on divers show dramatic improvement over conventional wet suits. One photographer on the Resolute Bay dive stayed in the water for more than four hours in a Unisuit.

There are also heated suits which use electric coils of low voltage to maintain a layer of warm air between the diver's body and the suit. Another method is pumping hot water inside the diving dress.

Exhalations account for much of the body heat loss. Since compressed air enters the lungs at temperatures below body temperature, it absorbs heat inside the body and carries it outside the body. With heliox, lungs cool even faster because the caloric capacity of helium is greater than that of nitrogen. Rebreathing systems have been used to try to reduce respiration heat loss.

Treasure Hunt

Reports of $4 million in cash and jewels have lured many divers to the bottom of the Atlantic off Nantucket Island, where lies the hulk of the luxury liner *Andrea Doria*.

Wrecks have long held a lure for divers, not only for the possible treasure, but also for the eerie scene a wreck makes and the strange marine creatures it attracts. In reality, not that much has ever been recovered because many ships that sunk were immediately stripped by their crews. Spanish sailors who manned ships laden with Indian treasures put buckets over their heads as they dived to salvage their cargo. In other cases, the rich treasures on the ships are part of the legend, perhaps originated for the benefit of an insurance company. But just enough has been recovered to keep such tales alive.

Diving on wrecks presents special problems: jagged edges and rotting timbers are common. Sunken ships deteriorate at a rate that varies with depth, materials used in their construction, exposure to pounding sea, chemical composition of the sediments on the bottom, and the type of boring and tunneling organisms—like shipworms—that live there. A ship may be unrecognizable in a matter of months after sinking or may remain almost intact for hundreds of years.

Bubbles from an aqualung can be dangerous, as divers who went down to the *Andrea Doria* found out. The ship, which sank in 1956, was in such dilapidated shape 17 years later that one diver said he felt that bubbles from his exhaust could shake loose several tons of wooden bulkheads which were hanging precariously from wires.

The *Andrea Doria* offers a good illustration of the difficulties involved with wreck diving. Ship manuals described the vessel as having been constructed with steel bulkheads covered with wood. But when the divers got there, they found that the whole interior of the wreck was made of false wooden bulkheads—in a badly deteriorated state. The danger of moving around in such a wreck ruined their entire salvage mission.

Although the *Andrea Doria* divers recovered only a crystal bottle of perfume and four silver plates, other more or less successful operations will certainly be undertaken.

The **Andrea Doria,** *a once-proud luxury liner, has been taken over by the animals of the sea. The reported $4 million in cash and jewels locked in her safes has also attracted man to the scene.*

In the 10,000 years that man has been sailing the seas, countless numbers of ships have gone to the bottom. **Studying the wreckage of ships** *provides us with knowledge about past civilizations.*

Diving into the Past

The storms and high seas that generations of sailors have cursed, fought, and sometimes succumbed to have provided us with historical and archaeological treasures. Man has been sailing the seven seas, or at least some of them, for nearly 10,000 years and the number of ships that have sunk are untold. True, the ancient wooden vessels would seem to be highly susceptible to the forces of destruction in the sea—the rot of the timbers, the drilling gnawing organisms, and the relentless pushing and pulling of the water itself. But most of these ships were rapidly covered by sediment, which protected them against further decay. Enough sunken ships, or their cargoes, have been found to spur archaeologists and anthropologists into increasing their efforts to recover historical evidence buried beneath the sea.

Diving has been the major tool of a science which is still in infancy—naval archaeology. Very little was known about the details and techniques of shipbuilding in antiquity be-

fore entire vessels, 2000 years old, were excavated by divers. Underwater exploration also includes sunken harbors and even whole cities. In the last 3000 years many towns and harbors in the Mediterranean and the Near East have become submerged.

It is only since the advent of the aqualung that archaeologists themselves have taken to the water. Prior to that they had to content themselves with digging in Egypt's Valley of the Kings, Italy's Etruscan ruins, Greece's numerous sites, and chance finds elsewhere. As far as marine archaeologists went, they had to be satisfied with hauling up artifacts with a net or sending down native divers or hard-hat sponge divers to bring up what they could. Now the archaeologists can dive under and explore firsthand the formations that might indicate a Phoenician trading craft, a Roman warship, a fishing boat of unknown origin, vessels of the Spanish Armada, or a sunken harbor.

The Mediterranean is a veritable graveyard of ships, not only because of its treacherous storms, but also because the sea has been

The development of the **aqualung** *in 1943 allowed archaeologists themselves to take to the water and lead the search for underwater treasure that was lost at sea centuries ago.*

traversed for so many centuries over routes so well known. And the virtual absence of tides and relatively weak currents in the Mediterranean help protect the wrecks.

The problems that plague marine archaeologists are similar to those that trouble his terrestrial brethern—getting the necessary permits from the bureaucrats who control offshore waters; beating the amateur treasure hunters to a site before they spoil it, plunder it, or just throw it into confusion; and finally, excavating the site by digging

The underwater camp of an archaeological expedition bustles with activity. Modern equipment and techniques have helped create a new science—naval archaeology.

through the encrusting creatures and layers of sediment in a scientific manner so that as much information as possible is recorded.

The difficulties are great and the cost of such operations is enormous, but the rewards are there too. Diving archaeologists are in the process of throwing a new light into the past.

Chapter II. The Call of the Sea

Today the seashores attract human populations as magnets draw iron filings. Most of the largest cities develop around harbors; major industries move to the coastlines; hordes of tourists invade shores and islands during the holiday seasons; millions of private boats and recently hundreds of thousands of diving outfits have been sold to people who have often moderate incomes. This "call of the sea" is, in fact, very recent.

Early man was more of a hunter than a fisherman. Even the islanders of the Pacific who had to extract their food from the ocean were fearful of the sea, and although in each generation a few young heroes ventured under the surface, it was only the demand from the Occident that induced Polynesians to develop the pearl-diving techniques.

It has been speculated that man's evolutionary roots had been matured in ocean water and that his recent return to the sea would

---

"The 'call of the sea' is recent. Early man was more of a hunter than a fisherman."

---

be just as natural as that of the sea mammals a few million years ago. Nothing can substantiate such a romantic statement. After all, the sea was not really very inviting to men. Storms took a heavy toll among sailors. Typhoons and hurricanes came from the open ocean and were included in the overall awe and suspicion. The water itself was cold and infested with stinging and biting creatures. Jonah may never have been swallowed by whale, but enough men found their ways into the stomachs of sharks to justify man's overall fear of the sea.

Much respect was earned by fishermen who did daily battle with the sea and by the navigators who dared traverse the ocean. But the superheroes were the divers like Gilgamesh, who 4000 years ago plunged to the bottom of the sea looking for a weed that would provide eternal youth, and Glaucos, the Greek mortal who became a god after eating a weed and diving into the ocean. Gods and animals, apparently, were the only ones at home underwater.

This, however, did not prevent a small number of human beings from diving. The Polynesian pearl divers and Greek sponge divers all have their secrets, legends, stories, and myths. The best-chronicled achievements are those of the Mediterranean, where the Greeks have been gathering sponges since immemorial times. Aristotle described the value of sponges to soldiers who used them to cushion heavy armor. Sponges were also used in the field as emergency dressings for wounds because of their ability to absorb liquids, and the warriors carried water-soaked sponges as canteens. Eventually early sponge divers provided not only indirect aid to the military, but took an active part in the wars by spying on enemy shoreline defenses, sabotaging ships, and transporting supplies through blockades.

It is only in the last few decades that science has exposed the myths of the ocean and that masks, fins, suits, and breathing apparatus have made it possible for modern man to answer the call of the sea and to enjoy her wonders in thrilling weightlessness.

*Modern man is answering **the call of the sea** in increasingly large numbers. Since time immemorial, the beauty and serenity of the underwater world and the thrill of weightlessness have drawn men to venture into the seas.*

ōan. Stradanus invent.   *Corn. Galle sculp.*

Down for Pearl

Sixteenth-century divers retrieve red coral from the bottom of the Mediterranean Sea. Sometimes wars were fought over the right to harvest coral.

Pearls are secreted by a few species of shellfish around the larvae of an intestinal parasite of rays. Rays become infected in eating and crushing the oyster and the pearl . . . thus releasing the parasite in their intestines. Most pearls are very small and irregular; valuable ones are exceedingly rare. In ancient times potentates (and more recently western jewelers) created a demand that motivated naked divers in the Red Sea, the Persian Gulf, and the South Pacific to risk their lives in collecting thousands of oysters to find perhaps one good pearl. When the Japanese suc-

ceeded in growing natural pearls by artificial farming, and when the salaries of the divers went up, pearl diving was no longer profitable and the hunt was carried on for the commercially valuable mother-of-pearl, which lines the shells and was used for buttons, earrings, knife handles, and ashtrays. Today the oyster beds of the Pacific have been depleted to such an extent that they are protected by law and diving is permitted only every other year on a small scale. In the Persian Gulf some of the beds are the personal property of local sheiks and are protected year-round by their armed guards.

Another rare natural treasure from the ocean is the "coral of the jewelers," a small tree of beautiful red, pink, or white limestone, not to be mistaken for the reef-building common corals. It used to be found in caves, but its rate of growth is so slow that it has been depleted in shallower depths and can today be obtained only in deep water by fully equipped divers or by submarines.

Easily the most romantic of the traditional divers are the pearl divers of Tuamotu Archipelago in the South Pacific. The men of the coral islands and atolls put aside their work on copra production from March to June and supplement their income by diving for large oysters, which must exceed five inches in breadth. The only equipment some of the men use are goggles, which help them find their lime-encrusted quarry.

The divers begin their ritual with a prayer asking protection from sharks and moray eels. Then they hyperventilate, usually emitting a low whistle when they exhale the quick, deep breaths. Then the diver wraps his toes around a rope with a weight at one end. When the weight is thrown overboard, the diver is pulled along, feet first, with little or no effort on his part. His only tools are a thick glove, which protects his hand from cuts and scrapes, and a collecting basket.

The diver surfaces on his own. If the area is productive, he may make as many as a dozen dives before moving on. He averages 40 or more dives a day, ranging from 100 to 140 feet down to collect 150 to 200 shells, and experiences severe physiological stresses.

A modern diver escorts a basket of red coral to the surface. Its scarcity has made it valuable—jewelers have paid $300 a pound for uncut coral.

Women: The Perfect Divers

Japanese women are among the most liberated when it comes to the sea, for women are better insulated against the effects of cold water. The loss of heat from the body is one of the biggest problems divers must overcome, for it restricts the length of time and the amount of work they can achieve.

The ama divers have been practicing their skills in Korea and Japan for at least 1500 years, and while they may have been pearl divers at one time, the 30,000 or so practitioners today dive almost exclusively for food, gathering shellfish and edible seaweed from depths of 20 feet all the way down to 100 feet. In times past, both sexes engaged in diving, but now there are few males prac-

ticing an art dominated by women. Because women have additional layers of fat beneath their skin, men are mostly relegated to the role of assisting the women divers by manning boats at the surface.

The ama women dive in summer when water temperatures may reach 80° F. in the Yellow Sea, the Sea of Japan, and the Pacific Ocean off Honshu and Shikoku. The Korean women also dive in winter, when water temperatures may drop to 50° F. at the surface. They wear only a loincloth and, in the last century or so, goggles or a face mask. In winter they wear a cotton swimsuit.

More important perhaps than the physiological difference, however, is the training the women receive. It may start when they are

Ama divers have been practicing their art for 1500 years. While they may have originally dived for pearls, today they dive almost exclusively for food.

girls of 11 or 12 and they may continue diving until they reach their 60s. Pregnancy and child care are no deterrents, for these women dive until the day they give birth and resume work shortly afterward, nursing their babies while they rest between descents.

The women usually hyperventilate themselves for five or ten seconds before a dive by taking deep breaths very rapidly. Then, just before plunging, they take a final deep breath, but not quite filling their lungs to capacity. On shallow dives, they may go 15 or 20 feet down, staying for about 30 seconds, with half the time spent actually gathering food from the bottom. When the amas have an assistant at the surface, they use a rope to help in ascending. Their dives then reach depths of 60 to 80 feet, and they stay

below for about a minute, spending half the time gathering food into a net around their waist. In either case, the women spend about 15 minutes out of every hour on the bottom, with another 15 minutes spent going up and down. The other time is spent resting at the surface between dives, with the rest periods twice as long between the deeper dives.

In Tierra del Fuego, at the "uttermost end of the earth," the women from the Yahgan Indians, today practically extinct, used to do all the diving of the tribes, descending completely naked in waters averaging 42° F. in order to collect clams and crabs. Diving requires very little muscular strength but great litheness of the body and great resistance to cold. Women thus should be the perfect divers, today as ever.

Ama women dive throughout the year. Their only protection in winter is a cotton swimsuit. Only recently have they used goggles or face masks.

Diving Warriors

The earliest dives made by men, or their anthropoid ancestors, were probably in pursuit of food. They first collected clams or shells at low tide, then stepped into the sea and ventured progressively deeper. Later, shells were also used for decorations. Such artifacts are among the oldest recovered by archaeologists. As early as 4500 B.C. divers were collecting mother-of-pearl as proven by the ornaments on the walls of Bismaya.

When historians began recording man's underwater feats, they often told of wartime activity. This does not mean that diving was necessarily a military art. More likely, the divers were recruited for their skills, which had been developed in their day-to-day activities. The historians are at fault for relating only tales of men fighting rather than of men in pursuit of peacetime activities.

Who were the first military divers? We don't know. It was falsely stated that Homer had mentioned in the *Iliad* their use in the

*Divers ride a **human torpedo** (above left), a submersible craft built during World War II. The two-man crew, carried outside the pressurized hull, wears **oxygen rebreathers** (above right) so that gas bubbles can't reveal them to the enemy.*

Trojan War some 32 centuries ago. In fact, he only made a poetic allusion to the attitude of a man jumping headfirst in the water.

By the fifth century B.C., two famous divers from antiquity, Scyllias from Sione and his daughter Cyana, went diving to cut the ropes from the anchors of the warships of the Persian king Xerxes. During a terrible storm the ships ran aground and sank. Later Scyllias and Cyana dove again to plunder the wrecks. During this same period the Greek historian Thucydides relates how Spartans used divers to ferry supplies past Athenian ships blockading the island of Sfaktiria. The Athenians eventually caught on to the idea and later employed divers themselves. The Athenians were attacking Syracuse in Sicily and their landing barges were stopped by submarine barriers of wood the Syracusans had erected. Underwater swimmers dismantled the barriers and allowed the Athenians to gain the upper hand temporarily.

One of the most spectacular uses of diving soldiers was during Alexander the Great's siege of the island stronghold of Tyre in 332 B.C. The divers were used to destroy the Phoenicians' submerged boom defenses, and Alexander reportedly watched the operation by descending in a glass barrel or diving bell. One of the oldest depictions of the scene, and there have been many different interpretations over the centuries, dates from thirteenth-century France. It shows Alexander's submerged conveyance illuminated by two candles which cast their light on various marine creatures, including an animal so large it took three days for it to pass by even though it was supposed to be traveling at the speed of lightning.

*The advantage of using **human torpedoes** is that the divers' hands are free to cut through any barriers protecting a harbor. They can then plant explosives on ships in shallow water, where the blast will be driven upward, through the ship's hull.*

Chapter III. Man and Machine

Man returning to the sea faces formidable difficulties. Water is cold and exhausts a diver insidiously. Water cannot be breathed, and increase in water pressure with depth makes it hard to force the indispensable air down to the diver. Lesser problems—darkness, turbidity, fear—may become insuperable for the ignorant or the coward.

Man, however, is a technological animal. He has devised many contraptions to supplement his limited ability to dive and work underwater. Man's efforts were spurred by potential economic gain in salvaging sunken ships or laying waste an enemy fleet. At that time the ultimate goal was to perform all these activities without getting wet while breathing air at atmospheric pressure.

Relatively few advances in diving machines were made before the sixteenth century

---

"The alternative was to dive within armor, or as a marine creature with the help of instruments."

---

when Leonardo da Vinci conceived a diving machine, perhaps a submarine. But Leonardo never revealed his plans, and they were kept secret, he said, "on account of the evil nature of men who practice assassination at the bottom of the sea."

The dreams, the ingenuity, the inventive drives followed two avenues: diving within an armor, completely isolated from water and from the effects of pressure; or, on the contrary, diving as a marine creature, even if to do so the human body had to be helped by various instruments. The first line of thought has today produced the bathyscaphe; the second has turned out the modern oceanauts. For a long time safety was thought to be associated with lines, chains, or links to the accompanying surface vessel. Now we know by experience, often tragic, that a self-contained vessel or diving apparatus enjoys safety as well as freedom of motion and efficiency.

Two main factors have constantly slowed down progress on pressurized and pressure-proof diving developments: the ignorance or the misunderstanding of the properties of the environment to be conquered and the lack of mechanical or physiological technologies. Accordingly the road has been long, the mistakes were naive or dramatic, and the penetration of the ocean was achieved a long time after it could have been done.

Obviously the diving armor will always be the only tool with which human beings will ever be able to reach the abyssal plains and the trenches, provinces that account for 90 percent of the sea floor. The deepest-known area of the oceans, the Marianas Trench, has been conquered by the bathyscaphe *Trieste* a few years after Tenzing and Hillary planted their flag on top of Mount Everest, but we are still in infancy when we build deep vehicles. Better materials to resist pressure, optical and acoustic improvements, greater maneuverability, manipulators as skillful as fingers, energy sources—in each of these fields, we need revolutionary new findings. Although nothing will ever completely replace the presence of man in the machine, the development of automated, unmanned, deep vehicles would be very helpful.

*In reality, water pressure would not allow this **imaginary diver** to move his arms in exposed flexible sleeves or draw air from the surface.*

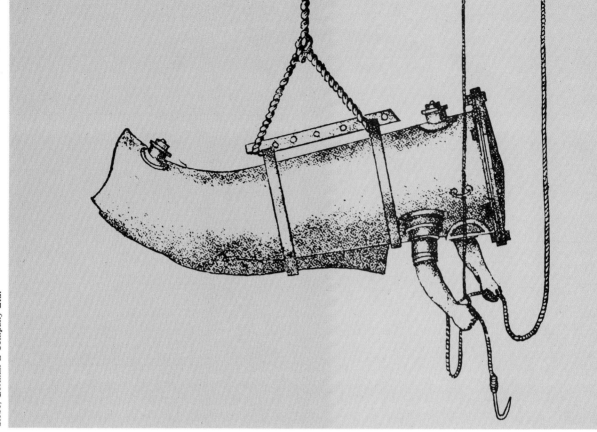

Siebe, Gorman & Company Ltd.

*In 1715 John Lethbridge invented **an illogical armored diving apparatus.** He claimed to have descended ten fathoms in it hundreds of times.*

Wishful Thinking

When early inventors decided to develop some sort of machine that would enable them to dive and at the same time stay dry and reasonably comfortable, avoid the effects of hydrostatic pressure, and be able to observe and act efficiently, they were unable to depart completely from the initial concept of the naked diver. Most of the designs, even those of Leonardo da Vinci, were as fanciful as they were unworkable, as impractical as they were contrary to the laws of nature.

Early inventors had only elementary technology available and very vague physical or physiological data. Such ignorance kept their imagination quite free from reality, but reality took its revenge when testings were attempted. The armor concept was the result of the long series of unsuccessful tests made with diving gear before the first use of the air pump by John Smeaton in 1788.

One of the first such attempts was a leather case designed in 1715 by John Lethbridge, who was interested in developing a vehicle to recover wrecks lost at sea. Here are extracts from his description of the machine: "It is made of wainscot, perfectly round, about six feet in length, about two feet and a half diameter at the head . . . and contains about 30 gallons; it is hoop'd with iron hoops without and within, to guard against pressure; there are two holes for the arms; and a glass to look through which is fixed in the

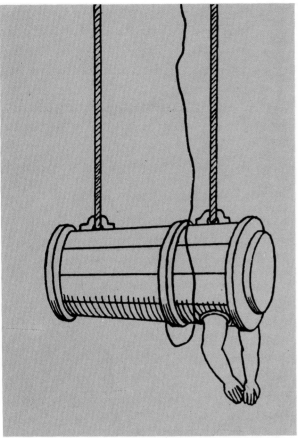

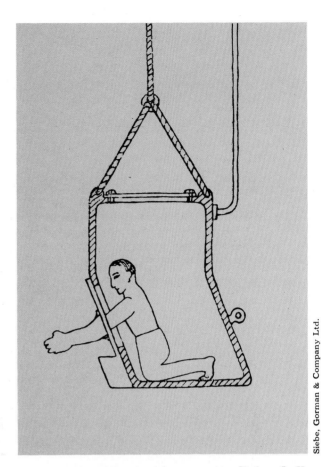

Siebe, Gorman & Company Ltd.

Siebe, Gorman & Company Ltd.

Rowe's diving machine was similar in design to Lethbridge's; armholes made both objects unrealistic for use at any depth below a few feet.

In 1903 John Watson patented this diving bell, which resembled Lethbridge's apparatus. Hands protrude through watertight sleeves.

bottom part to be in a direct line with the eye; two air-holes, upon the upper part, into one of which air is conveyed by a pair of bellows, both of which are stopped with plugs, immediately before going down to the bottom. There's a large rope by which it's let down. . . . I go with my feet forward, and when my arms are got thru' the holes, then the head is fastened with screws. It requires 500 weight to sink it. . . . I lie straight on my breast, all the time I am in the engine, which hath many times been more than 6 hours, being, frequently, refreshed upon the surface by a pair of bellows. . . . I have been 10 fathoms deep many a hundred times. . . ."

Unfortunately it is obvious today that only a completely enclosed armor allows the diver

to remain at atmospheric pressure. Any flexible part, as the flesh of the diver's arms, is inexorably forced in by water pressure as soon as depth reaches a few feet and the diving attempt is soon a disaster.

A very similar design was built shortly after Lethbridge by a Captain Rowe. It also had "armholes" and so was unrealistic.

The same technological problems were faced by the first builders of submarine vehicles—Drebbel in 1620 and de Son in 1653. Their submersibles were built of wood and could not dive to any significant depth. Their propulsion systems, which used direct or indirect manpower, did not give them an appreciable range of operation.

The Human Lobster

The quest for a device that would allow a man to work underwater and still breathe air at atmospheric pressure could theoretically be solved by an articulated armored diving dress. Simple enough, then. The diving apparatus had to be rigid and heavy enough

The Carmagnole diving armor was built in 1882 and incorporated 22 ball-and-socket joints into its construction to give the diver more mobility. These joints were an improvement over accordion types, which shrank and stiffened under pressure.

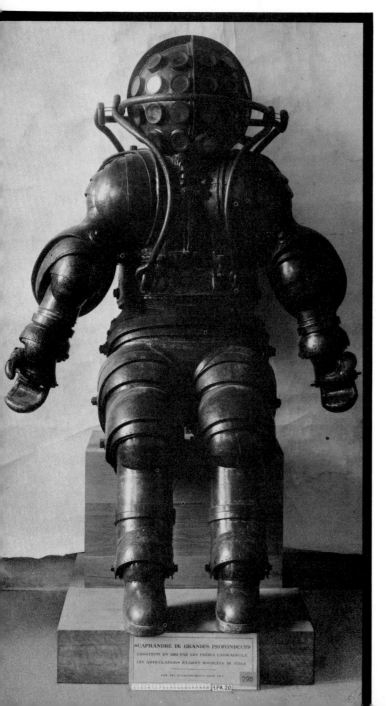

SCAPHANDRE DE GRANDES PROFONDEURS
CONSTRUIT EN 1882 PAR LES FRÈRES CARMAGNOLE
LES ARTICULATIONS ÉTAIENT DOUBLÉES DE TOILE

795

1 PA 20

to sink and to withstand the increased pressure beneath the surface; it had to provide for scrubbing the air inside to keep it suitable for breathing; it had to entirely encase the diver and yet allow mobility in the legs, arms, and hands. Early designs of such underwater suits called for shoulder, elbow, wrist, knee, and other flexible joints, made of leather accordion pleats stiffened by rings, or some other sort of corrugated material, resembling a bellows. Unfortunately such materials shrink and stiffen under pressure.

It wasn't until 1882 that two Frenchmen, the Carmagnole brothers, patented a diving dress made of metal which had 22 ball-and-socket joints. In these heavy closed suits, the rigid metal offered protection from hydrostatic pressure but restricted the mobility of the diver. The ball-and-socket concept was an improvement over the bellows joint because it was not as adversely affected by pressure. The Carmagnoles used treated linen to line the joints in order to keep them waterproof, but with 22 joints it must have been difficult to keep them all in top condition. Perhaps the most distinguishing feature of the Carmagnole suit was the helmet, dotted with a quantity of small windows spaced about the same distance apart as the eyes of a diver.

Shortly after the Carmagnole suit appeared, Englishman William Carey sketched a pattern for a diving dress that used roller and ball bearings to ease the action in the joints. This dealt partly with the problem of restricted movement in the heavily armored deep-diving dress; other inventors worked in vain on more flexible suits stiffened with spiral wires in an effort to achieve mobility without the rigidness of the all-metal suit. No matter what the improvements were in the critical areas of the articulations, 200 feet deep was virtually a barrier until the early part of this century.

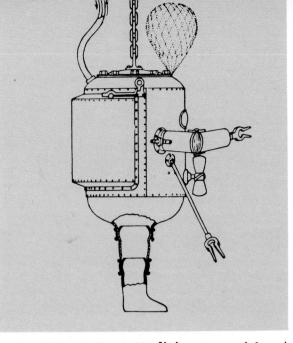

Siebe, Gorman & Company Ltd.

*L. D. Phillips designed this **diving armor** (above) in 1865. It was a short thick metal cylinder with domed ends. Limbs were fitted with ball-and-socket joints and hands remained inside arms. Several of its features appear in modern diving carapaces.*

In 1913, German designers Neufeldt and Kuhnke patented a rigid suit which combined ball bearings and ball-and-socket joints to increase flexibility. This impressive suit was put into production in 1920. Other fairly successful deep-diving armored dress were designed by Galeazzi and by J. S. Peress, who used a liquid to seal the joints.

All these contraptions had to make a choice or a compromise between a few joints, which gave little freedom of movement, or a multitude of fragile articulations, which allowed greater movement. All joints became very stiff deeper than 300 feet and the weight and the cumbersomeness of such gear make it difficult and dangerous to bring them into operation if the sea is not perfectly flat.

The first successful salvage operation in very deep water was performed by divers on the S.S. *Artiglio* in 1931. They recovered 95 percent of the gold carried by the S.S. *Egypt*, which had sunk in 400 feet of water. The *Artiglio* crew tried first to use the Neufeldt and Kuhnke rigid suit; they quickly abandoned it in favor of a Galeazzi chamber, which directed manipulators lowered from above.

A human is handicapped when he tries to shield himself against pressure by becoming an invertebrate "lobster" in articulated armor. His mobility is reduced and ultimately he is put at the mercy of the sea.

*In 1920 Benjamin Franklin Leavitt proposed to use his **diving dress** (below) to salvage the Lusitania resting in over 300 feet of water. However, water pressure at that depth would collapse the suit's flexible limbs. The salvage was never attempted.*

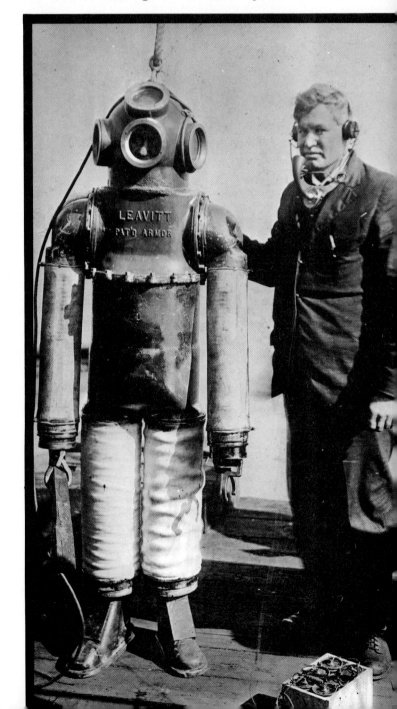

Submerging Chambers

A deep-diving armored dress is a one-man articulated submersible, with a tightly bolted upper entrance cap. The upper half contained an observation window and articulated arms, while the lower portion had movable legs. This basic design was used tentatively in the salvage operations of the *Egypt* but was abandoned in the 1930s. Standard diving suits are more efficient down to 300 feet; at lower depths the rigid articulated gears are made ineffective because pressure stiffens the joints.

The famous master diver Gianni on board the *Artiglio* was the first to understand that he could carry out the difficult salvage of *Egypt*'s gold much more efficiently in a submerged chamber from which he could moni-

*Man has always dreamed of venturing below the waves. This **early diving sphere** was designed by Nicolo Tartaglia in 1554.*

Siebe, Gorman & Company Ltd.

*The steel claws of a **scissors grab** lift the captain's safe, weighing half a ton, from the wreck of the Egypt, submerged in about 70 fathoms of water. A man in a diving chamber below directed the grab into place for its catch.*

Siebe, Gorman & Company Ltd.

***Dangling at a depth** of nearly 400 feet, an observer in this chamber directed the salvage of the Egypt. The treasure chamber was uncovered, and almost all of the five tons of gold and two tons of silver that had gone to the bottom was retrieved.*

tor the handling of dynamite and of mechanical claws lowered from the surface.

Such chambers are tube-shaped, with the cylindrical shaft large enough to hold a man standing. Because there was no need for flexible limbs, these chambers were much less expensive to construct and use and were far more reliable than diving suits. Underwater observation chambers had venerable ancestors, not counting Alexander the Great—in 1865 Ernest Bazin tested one of his inventions in 250 feet of water, and in 1890 Balsamello was lowered in his sphere down to 430 feet. The Italian engineer Roberto Galeazzi has constructed both articulated armored diving dress to be used at a maximum depth of 800 feet and observation chambers effective at depths below 2000 feet, where the pressure exerted by the water exceeds 1000 pounds per square inch.

As much of an improvement as the cylindrical shape is over the limbed diving dress, it does have limitations compared with a spherical shape. Galeazzi knew this, and his roughly cylindrical design is in fact made of many short spherical sections welded on top of each other. William Beebe showed the superiority of the sphere shape in the 1930s when he used a bathysphere to exceed depths of 3000 feet. Beebe's observation chamber had three windows for vision and was tethered to a mother ship at the surface by means of a steel cable, containing electric and telephone lines. The bathysphere had its own self-contained breathing system of oxygen; it used soda lime to absorb exhaled carbon dioxide and calcium chloride to remove moisture from the enclosed atmosphere. Dr. Beebe, a naturalist, made several descents in the bathysphere, and in 1934, with Otis Barton, reached a depth of 3036 feet in the Atlantic off Bermuda. In 1949 Barton alone reached 4540 feet in his benthoscope.

Diving in midwater in a device suspended under a surface vessel is difficult when the mother ship is rocked by surface waves. In this situation it is dangerous to work close to the bottom or to a shipwreck; the vertical forces applied to the cable by the rolling and pitching mother ship are transmitted to the chamber and in some cases are amplified by the elasticity of the cable. To overcome this effect, the Japanese "Kuroshio" turrets have self-contained propulsion and buoyancy mechanisms electrically fed by the power cable from the accompanying vessel to move and stabilize the underwater capsule.

With the development of closed-circuit television, some unmanned tethered vehicles such as CURV or Telenaute have been successfully used on underwater missions.

The sphere is superior in structural design to cylindrical diving chambers. In 1934 Beebe and Barton descended 3036 feet in this **bathysphere.**

Cutting the Cord

Puces (above, below right) are modern minisubs with great research potential. They can carry a man down 1600 feet, and external manipulators give him artificial arms outside the craft.

As long as men in diving suits and observation chambers were still leashed to ships at the surface, their freedom of movement beneath the water was greatly restricted. The ideal had not been totally accomplished.

Submarines had long been a goal of designers, and as early as the seventeenth century, inventors like Cornelis Drebbel and de Son attempted to build covered boats to move underwater. These wooden vessels were propelled by oars or man-powered paddle wheels and were barely submerged. They were generally tested in rivers, and if they moved at all, it was probably as much the result of the current as of the rowers. Many early submarines were designed for warfare, but they were far more dangerous to their crews than they were to the enemy.

The credit for designing the first true mobile submarine usually goes to David Bushnell, who designed an egg-shaped boat that was powered by means of manually operated cranks attached to screw-type propellers. As Bushnell described it: "An oar formed upon the principle of the screw was fixed in the forepart of the vessel; its axes entered the vessel forward, but being turned the other way, rowed it backwards." Bushnell's submarine, the *Turtle*, was used in the Revolutionary War to move beneath British warships and plant explosives.

Robert Fulton, of steamboat fame, was commissioned by Napoleon to design a submarine. His vessel was successfully operated beneath the surface, but it was really not all that submersible since it had to stay very near the surface so that the operator could guide the craft's course. The development of submarines was almost inextricably linked with warfare, and for nearly 200 years the concept of self-contained mobile observation chambers lay fallow.

Recently, however, we have been using minisubs launched from the *Calypso*. We call them puces, French for fleas. These minisubs give the diving pilot artificial arms through external manipulators and visual memory through cameras mounted on their backs.

The **Turtle** *was the first true mobile submarine and was used against Britain in the Revolutionary War.*

The pilots are protected from the high pressures of the underwater environment, even down to depths of 1600 feet, and are able to breathe air at atmospheric pressure. Perhaps more importantly is the ability of these vehicles to be used in tandem, one being able to rescue the other if such a need arose.

The puces are smaller, faster, lighter descendants of our diving saucer, *Denise*, launched in 1959. *Denise* was a two-man hydrojet craft, good for depths of 1150 feet. It was the first submersible small enough to be carried aboard *Calypso*.

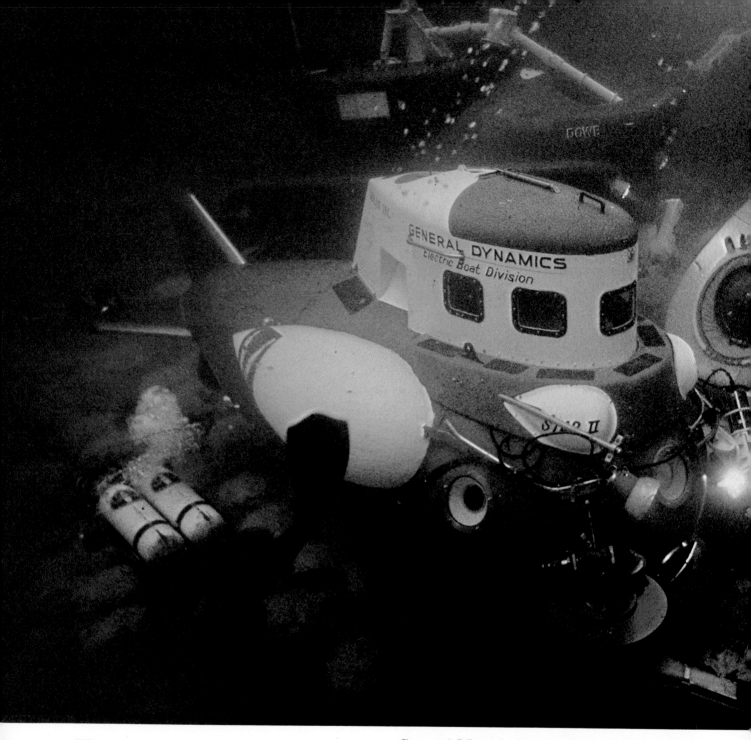

The Submersible Boom

By the late 1960s many self-contained exploration subs had been built in France and in the United States. In order to demonstrate their capabilities, in the spring of 1969 a jamboree of submarine vehicles was organized off the island of Catalina. During several days, seven subs—Lockheed's *Deepquest*, North American Rockwell's *Beaver*, General Motor's *Dowb*, General Dynamic's *Star II*, General Oceanographic's *Nekton*, and *Calypso*'s two one-man puces—were discussed and compared. They also dived together and were filmed. It was a memorable occasion, although some remarkable submersibles were absent, among them Wood's Hole's *Alvin*, Reynold's *Aluminaut*, Grumman's *Franklin*, Westinghouse's *Deepstar*, and *Calypso*'s diving saucer.

of divers, and with as much maneuverability as possible. They not only increased the depth range, but could also stay much longer than men submitted to pressure and cold.

Each sub was designed either for different tasks or along various concepts of what the challenges of deep diving were. The largest is the *Aluminaut,* with an aluminum hull 51 feet in length, cigar-shaped and capable of cruising at more than four knots. It carries a crew of four down to 15,000 feet and has an endurance of 32 hours.

Basic choices have to be made at the conceptual stage of an exploration sub. The need for speed suggests an elongated streamlined shape, but maneuverability is maximum with a saucer design. If the vehicle is hoisted on board the tender vessel, a simple hatch for the entrance of the crew on board is sufficient; but if the submersible is too heavy to be handled by a crane in a rough sea, then it will be towed by the mother ship to the diving area, and the entrance hatch must be protected by a cumbersome conning tower. Most builders aim at a speed of three knots, but maximum good exploration speed is one knot. Visibility is best obtained by wide angle portholes, but some have built the entire hull in transparent acrylic plastic *(Deep-View, Sea-Link);* others use undersea periscopes *(Dowb);* and others use mainly television *(Aluminaut, Deep Quest).* Safety is a basic concern of the designers of both one-man subs and vehicles with a larger crew. Two men on board seems to be the best solution. Neutral buoyancy is most economically obtained by lightweight high-yield construction materials, but often it requires incompressible ballasts as additional floats.

All these incredible diving machines have one characteristic in common: they endeavor to make direct observation possible by men deep in the sea. Aqualung diving had demonstrated that scientists could better understand the problems they were studying if they could observe geological phenomena or animal behavior with their own eyes. The new vehicles intended to carry on such visual surveys at depths well below the range

Star II, puces, and Dowb congregate on the ocean floor. These diving machines have made it possible to study the sea through direct observation, at depths below the range of aqualung divers.

45

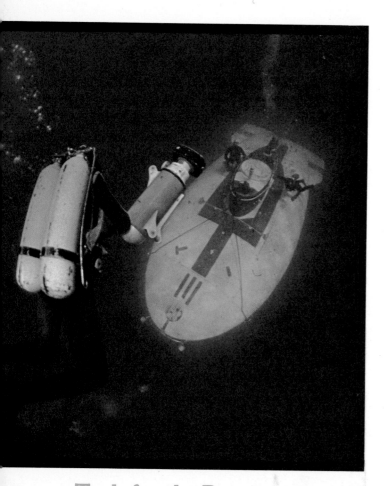

Even at lesser depths of 2000 feet, the vertical voyages down to the bottom and back to the surface are time and energy consuming. In order to minimize these problems, most crafts use hydrostatic forces: they are fitted before each dive with two jettisonable pig irons, heavy weights calibrated to sink the sub at good speed. When the pilot reaches the bottom, he drops the "descent weight," and his machine becomes practically neutrally buoyant; at the end of the dive, he drops the "ascent weight" and the sub floats back quite naturally to the surface. Some builders, obsessed by the tradition of conventional military submarines, use air ballasts that can be filled at will with water and emptied by pumps or compressed air. But the ballast system, while good for shallow crafts, is inefficient and dangerous at great depth. All submersibles have emergency weights, and in case of danger even the heavy propulsion batteries and the mercury normally used for trimming can be jettisoned.

Propulsion is most often obtained by watertight electric motors or by motors running in oil and coupled with propellers. Some crafts even are fitted with horizontal, vertical, and transversal props, working inside a protective tunnel to avoid fouling or damage. The diving saucer and *Calypso*'s one-man puces are propelled by hydrojets with rotating nozzles that can direct their thrust in practically any direction. Jet propulsion allows better streamlining, and nozzles are safer than propellers because they cannot be entangled; if they were, they are detachable at a low replacement cost. Unfortunately the efficiency of a small hydrojet is still about half that of a propeller, which limits the practical range of action. The electric batteries used are heavy; a gigantic improvement in performances of the exploration subs will come with lighter fuel cells.

Tools for the Bugs

The earlier subs, such as the diving saucer, were conceived to operate easily on the whole of the continental shelf and a little deeper. This goal was set because the shelfs of the world total a surface comparable to Asia and were considered as the richest area of the sea. But they are only 8 percent of the ocean floor, and oil and minerals have recently been discovered much deeper.

The next province to be explored is the continental rise (2 percent) and the abyssal plains (85 percent) at an average depth of about 13,500 feet! Leaving aside the family of undersea vessels called bathyscaphes, which have the greatest depth capabilities, the *Aluminaut* today is the only exploration sub capable of working in the abyss—it's a long way down and the pressure is great.

Instruments and tools are important factors of efficiency for undersea vehicles. Sonar telephones with the surface, sonar and echo sounders, gyrocompass, searchlights, and cameras are standard equipment. But practical manipulators, sampling and coring tools, and simple bottom-anchoring systems are still in infancy. Those that are available are either clumsy and expensive or very inefficient. Effective tools for the bugs are a must if the bugs are to know the success they deserve. After all, they are man's only hope to understand the deep-sea environment.

A diver with an underwater movie camera (left) swims toward the submersible Deep Quest.

Beaver IV *(above) has a mating skirt so that it can transfer divers to an underwater habitat.*

The submersible **Deepstar** *(right) is mounted with a stereo mapping camera and powerful strobe lights to photograph the sea bottom.*

Undersea Dirigible Balloons

Already famous for his daring, record-breaking balloon ascents, Auguste Piccard, a bold and colorful Swiss professor at the University of Brussels, decided he would set out and conquer the deep hydrosphere in the ocean. His first goal was the abyssal plains, 13,500 feet below the surface. A steel sphere, capable of withstanding pressure of almost 7000 pounds per square inch, would accommodate two men; it would be heavy, even in the water, and would have to be suspended or buoyed up. Piccard rightly decided that a cable, such as had been used by Beebe and Barton to connect the sphere to a surface vessel, would be too long, too heavy, and too dangerous. The balloon solution came naturally to his mind. There were two major changes: the light, but compressible gas would be replaced by a liquid lighter than water—gasoline; and propellers would turn his bathyscaphe (deep boat) into an undersea dirigible balloon.

A first project was abandoned before it was built; the second bathyscaphe, named *FNRS II,* made an unmanned test dive to 5000 feet in 1948 at Cape Verde islands; the third one, *FNRS III,* was built by the French navy and held a temporary record in 1953 in a successful trip to 13,000 feet with Commander Houot

Auguste Piccard, a Swiss balloonist, became a famed ocean explorer. His second bathyscaphe, **FNRS II** *(above), descended unmanned to 5000 feet in 1948.*

Trieste *(right), another Piccard-designed bathyscaphe, descended 35,800 feet. Its gondola can withstand pressures of seven tons per square inch.*

Built by the French navy, the **Archimede** *(below) went 30,600 feet down into the depths of the Kuril-Kamchatka Trench between Japan and Russia.*

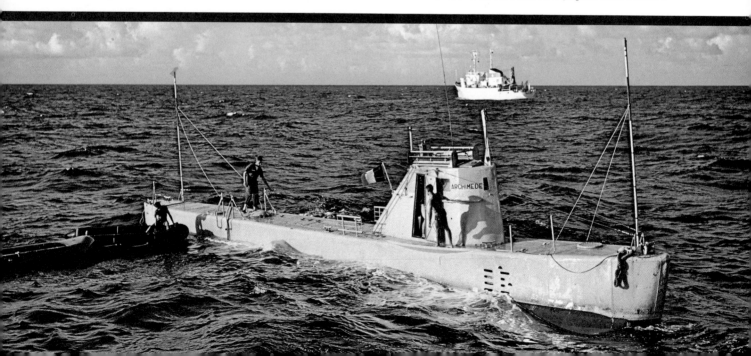

and engineer P.-H. Willm on board. In Italy Auguste Piccard was then building another bathyscaphe, *Trieste,* good for 20,000 feet. In 1958 the *Trieste* was acquired by the U.S. Navy; it was reinforced and the steel spherical gondola was changed for a stronger one, made in three sections by Krupp in Germany and able to resist the formidable pressures to be met at the bottom of the deepest trenches in the world—seven tons per square inch!

On January 23, 1960, the *Trieste,* manned by Lieutenant Don Walsh and Piccard's son Jacques, reached the bottom of the Marianas Trench, 35,800 feet deep, after 4 hours and 48 minutes of descent. As the *Trieste* landed, it stirred up the bottom sediment to such an extent that the ship was engulfed in a white cloud. This white material was found to be "diatomaceous ooze," the remains of one-celled plants called diatoms that had filtered down from the surface to the depths of the ocean. Walsh and Piccard stayed 20 minutes on the antisummit; they observed one shrimp and one flatfish (which proved that life existed in the deepest part of the ocean), and they came back to the surface, concluding the historic dive by a 3-hour-and-17-minute ascent. The Americans had conquered the Everest of the sea with a vehicle designed by a Swiss professor in Belgium, built in Italy and including a German sphere. The ocean was ironically demonstrating how effective international cooperation could be.

The *Trieste* is obsolete. Another bathyscaphe, stronger, faster, also capable of withstanding pressure in the deepest trenches, was built by the French navy. The *Archimède* explored the Kuril-Kamchatka Trench, 30,600 feet down, in 1967. It is now engaged in the Franco-American exploration project FAMOUS in the Mid-Atlantic Ridge.

49

Chapter IV. From Air to Water

The two most important fluids in the world, at least as far as life is concerned, are air and water. Indeed, life on our globe is divided on the basis of those creatures that live on land —in the air, or atmosphere—and those that dwell in the sea—or hydrosphere. These two fluids behave according to the same physical laws, but because their physical and chemical makeups are so different, the results differ greatly. For example, both air and water exert a pressure on the human body, but at sea level a column of air extending thousands of feet to the end of the atmosphere has a force of 14.7 pounds per square inch, a whole column of seawater only 33 feet high exerts exactly the same amount of pressure. These differences make swimming and diving what they are, for the hydrostatic pressure of water aids in buoyancy, pushing

---

> **"The two most important fluids, as far as life is concerned, are air and water."**

---

up on the body in direct proportion to the volume of water that is displaced.

It is important to remember that air is a mixture of gases, mostly oxygen and nitrogen with lesser amounts of argon, carbon dioxide, water vapor, hydrogen, helium, and neon. Within solids there is a rigid structure and a physical limit on the movement of the molecules; but in fluids molecules are freer and can move among one another. A liquid can dissolve another substance which uniformly diffuses into it. Eventually, however, a limit is attained. This occurs when the liquid reaches the saturation point. The quantity of material diffused depends upon the nature and temperature of the two substances in question. As a result, scientists have been able to determine tables of solubility for two elements at the same temperature.

Gas, the third basic state of matter, always diffuses more or less into the element with which it comes in contact. Even two different gases put together will act as though each were alone, and each will spread throughout the entire container holding them. Two gases of similar density will mix almost instantaneously, while two gases of radically different densities will mix much more slowly. Gases in a mixture exhibit a unique property, for each not only expands to fill the entire volume of the container, but each also exerts its own pressure on the container. In other words, each gas in a mixture of gases exerts a pressure that is independent of all other gases. The force exerted by each gas is called partial pressure, and the force exerted by the mixture—called total pressure—is the sum of all the partial pressures of the constituent gases. In air, for example, the atmosphere is composed largely of oxygen and nitrogen. The minute amounts of carbon dioxide, water vapor, and rare gases may be ignored for this purpose. Oxygen makes up about 21 percent of the air and nitrogen about 79 percent. If air were trapped in a container and all the oxygen removed, the pressure would drop from 14.7 pounds per square inch (psi) to 11.6 psi. And if the nitrogen were removed instead, the pressure would drop to 3.1 psi. Thus, the partial pressure of oxygen is 3.1 psi; that of nitrogen is 11.6 psi; and their total equals 14.7 psi, the absolute pressure of air at sea level.

Atmosphere and hydrosphere are two radically different environments into which Homo sapiens has ventured. To the diver, men above look like puppets.

As this **plastic parachute** moves to the surface, the air trapped under it increases in volume because the surrounding hydrostatic pressure lessens.

Pressure Points

The physical laws of pressure were first put together by Robert Boyle in the 1660s, and they applied to all gases. Other scientists, divers, and mountain climbers before Boyle may have had a working knowledge of the laws, but they failed to state them in a manner acceptable to the scientific community.

Boyle's law states that for a body of gas at a given temperature the volume is inversely proportional to the pressure. In other words,

the more pressure exerted on the gas in the container, the less volume there is; and the less pressure that is exerted, the greater the volume. For example, a balloon filled with helium rises in the atmosphere and keeps expanding as air pressure decreases until it bursts. Another demonstration is afforded by human lungs. A diver taking a deep breath before plunging into the sea would find his lungs being "squeezed" as he descends and as the water pressure increases on the air trapped inside the body. His chest subsides and his belly grows hollow as if he were exhaling. When he swims back to the surface, chest and belly come back to normal.

Solids, liquids, and gases are composed of free molecules that are attracted to each other and move at speeds increasing with rises in temperature. When water is heated in a container and boils, molecules escape and become gaseous water vapor. If the container were tightly closed and if the temperature rose high enough, the container may break with an explosive force. Gases will always expand to fill the container that holds them and, if the container is elastic like a balloon or a lung, will expand the container too. Gas in and of itself has no determined volume, so the pressure it applies on the sides of its container is proportional to the speed and number of molecules. Its pressure is called kinetic pressure.

All the molecules of the elements on the earth's surface are attracted vertically toward the center of the earth by the force of gravity. The atmosphere around the earth, for example, is trapped in no container, but is held around the earth strictly by the force of gravity. In a solid, the weight it exerts is the amount of gravitational attraction on the combined number of molecules in the object. In the case of a liquid the force of gravity determines hydrostatic pressure, measured by the weight of the column of liquid above it. So at the surface the hydrostatic pressure is zero. In incompressible liquids, such as water, the density is practically constant and the increase of pressure is regular and rapid as the depth increases.

We are so accustomed to the air around us that we do not always notice its density and pressure. But the density of air is very important in determining whether airplanes can lift off the ground. And the pressure of air is uniformly exerting a force on everything at sea level of one atmosphere, or 14.7 pounds per square inch (usually abbreviated psi). Rising from sea level, the column of air that weighs on us becomes shorter, and atmospheric pressure decreases, so that on top of a 16,500-foot mountain, it is only half as much as it is at sea level. Correspondingly, the atmospheric pressure increases at points on land that are below sea level.

Water is 800 times as dense as air; thus, when descending in water below sea level, the pressure increases very rapidly. At 33 feet below the surface, the pressure of the water itself is 14.7 psi, and since the air above the water is also exerting a force of 14.7 psi at the surface, they combine to put pressure on a fish or rock or human of "two atmospheres," which is 29.4 psi. Descending another 33 feet in the water adds another 14.7 psi of pressure; at 66 feet down, a diver is subjected to slightly more than 44 psi, or three times as much pressure as at the surface. If man, with all the air cavities inside his body, wants to reenter the sea, pressure is a force that must be dealt with.

Air that fills an inverted pail or human lungs at sea level is squeezed to half its volume 33 feet under water, where the pressure is twice as great as at the surface. At 66 feet, pressure is tripled and volume is one-third; at 99 feet, pressure is quadrupled and volume one-fourth as great. The lungs are not shown in proportion in the illustration because of their irregular, three-dimensional shape.

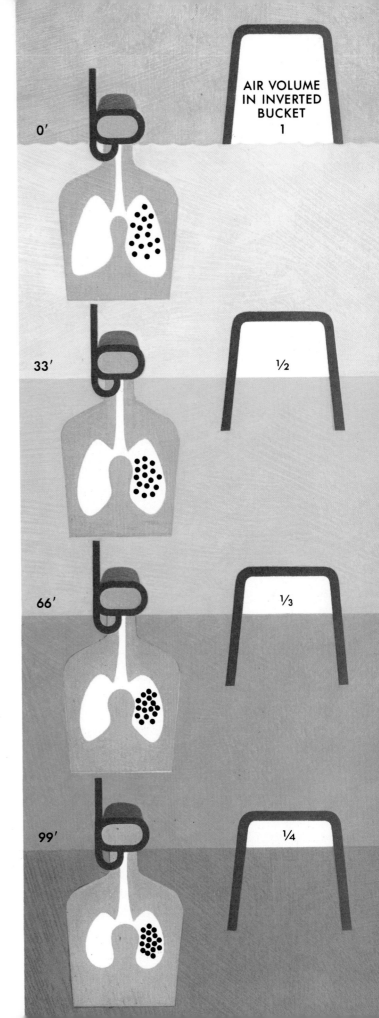

*As **rays of sunlight filter down** from above, Frederic Dumas seems to hover effortlessly over a sea floor strewn with jagged rocks.*

Strength of Water

The ancient Greek mathematician, physicist, and inventor Archimedes formulated a principle dealing with hydrostatic pressure, although he obviously didn't know all about it at the time. Archimedes probably thought he had just figured out why things float. He determined that an object immersed in a fluid is buoyed up by a force equal to the weight of the fluid displaced. In other words, if that submerged body weighs less per cubic inch than water it will float, but if its density is greater than water's it will sink. As an example, if a diver breaks his hammer, the head will sink and the handle will float. As water is practically incompressible, its density remains constant at all levels and the upward thrust of buoyancy does not depend on the depth at which solid objects or incompressible bodies are submerged. But in the case of a gas in a flexible container or other compressible object, its volume decreases as the pressure increases. And so its buoyancy correspondingly decreases with the volume. The flexible bag might be in equilibrium at some point and hang suspended in the water. This means that the buoyancy of the fluid is just strong enough to offset the gravitational attraction pulling the object down. But this equilibrium is unstable. If the object rises slightly, it swells and the buoyancy increases with the effect of pushing the object up even farther. The volume increases some more, and the chain reaction continues, since the buoyancy has become stronger than the object's weight and it will rise faster and faster as it expands until it will reach the surface where it will float. As might be expected, exactly the opposite effect is produced if the air bag in equilibrium descends somewhat. It will sink more and more rapidly as the volume—and thus the buoyancy—decreases, until it reaches the bottom.

The human body is made up mostly of solids, liquids, and air passages connected to the nose and to the mouth. Flesh and bones will change little in density as depth increases. The air passages and cavities, on the contrary, are elastic. A naked diver will float at the surface but below 30 or 40 feet he will sink because pressure has reduced the volume of gas within him and he has become less buoyant. One of the most important factors in determining a human body's buoyancy is the volume of air in the lungs. The difference between a full inhalation and a full exhalation is about 20 percent of a cubic foot, which translates into a change of about 12 pounds in buoyancy.

These figures and computations are based on an ideal theoretical fluid and there are differences between theory and practice. In air, for example, the thrust on a man is very slight. If the man weighs 150 pounds, the buoyancy thrust is about three ounces at sea level, and 1.5 ounces at an altitude of 16,500 feet. The same 150-pound man in fresh water will be subjected to an upward thrust of about 148 pounds, and he will sink slightly. But salt water is denser than fresh water, and in the ocean the 150-pound man will float as a result of a buoyancy thrust of about 155 pounds.

*As water and sediments are replaced by air from the diver's mouthpiece, the buoyancy of this **amphora** is increased, and it moves toward the surface.*

Natural Limits

One man was able to hold his breath, after ventilating his lungs for a long time with pure oxygen, for 15 minutes and 13 seconds, which is pretty close to what the dolphin can do. But without such oxygen hyperventilation, the record is only six minutes. In fact, exceptionally gifted naked divers can stay only a maximum of four minutes, and only in warm waters.

Another natural limit of the human body has been repeatedly tested: the depth to which a naked man can descend without any breathing aid, without permanent injury.

On August 18, 1973, the Italian diver Enzo Maiorca, from Syracuse, Sicily, who was 42

Enzo Maiorca is surrounded by his companions after diving to a record depth of 265 feet, unaided by breathing apparatus.

years old and the father of two, after eight minutes of hyperventilation, seized a 50-pound weight and sank rapidly down, off Portovenere. He wore no fins, no mask, no goggles, no breathing apparatus; he was only protected against cold by a foam rubber wet suit four millimeters thick. Maiorca reached 265 feet, grasped a control tag, dropped his weight, and then ascended rapidly. His record dive had lasted only 2 minutes 18 seconds. He was stressed, but recovered rapidly.

How many unknown natives have reached such depths? In December 1913 the Italian cruiser *Regina Margherita* lost its anchor in Greece. A local sponge diver, Haggi Statti, who was 35 years old and the father of four children, repeatedly dived naked to 265 feet and successfully recovered the anchor. He used a 30-pound weight to sink and claimed that he had already been deeper.

The ritual may vary, but in all very deep dives the diver hyperventilates before descent. Hyperventilating increases the amount of oxygen in the blood and in the muscles as it decreases the amount of carbon dioxide. Hyperventilation unquestionably allows divers to hold their breaths longer underwater. Ordinarily, the body signals when the organism is not getting enough oxygen. The brain cells are the most susceptible to be damaged by anoxia, and as a warning signal when the carbon dioxide level in the blood reaches a certain point, the body says, in effect, "take another breath." Unfortunately, as we know today, this reflex is inhibited by high partial pressures of oxygen, which happens in deep dives. Moreover, for the same reason, during extended deep dives, the percentage of oxygen in the lungs can drop to very low levels without becoming immediately dangerous, because the oxygen *partial pressure* is still high enough to be equal to or more than that which exists in our lungs at the surface. But during the ascent, this partial pressure decreases dramatically (by a factor of nine during an ascent from 265 feet) and becomes too small to regenerate the blood. As a result, the test diver may pass out before reaching the surface.

There is another interesting reaction the body performs automatically when a person, or any cold- or warm-blooded animal dives: it is the diving circulatory reflex; it automatically slows down the heartbeat to reduce oxygen consumption. This reduction of the heart rate is moderate in man, greater in sea mammals, and spectacular in the case of the marine iguana when all body circulation, except that to the brain, is stopped during a dive. At the same time, the skin vessels are constricted, the peripheral circulation is blocked, and the external layers of the body cool off and become a buffer zone, which helps keep the central temperature constant.

Maiorca, as he nears the surface after descending 265 feet. When a person or animal dives, the body automatically reduces heartbeat to conserve oxygen. In fact, the marine iguana can shut off all body circulation except that to the brain.

Chapter V. Air to Breathe

If diving was going to yield any practical results for human beings, man had to devise some method of staying underwater. He couldn't extract sufficient oxygen from the water itself the way fish do, and his breath-holding capacity was severely limited and nowhere near as efficient as that of his cousins, the dolphin or the whale.

In nature, some insects have been successful in solving the same problems by developing either a breathing tube extending to the surface or submerged containers of air filled by shuttling to and from the surface of a pool. The protractile appendix of the one-inch-long whip-tailed larva of the drone fly (*Eristalis tenax*) may extend several times the size of the creature and allows them to draw air as deep down as six inches. Bathing

---

"Humans were limited in diving operations before the invention of the air pump."

---

elephants may walk on the bottom until they are submerged, and then they use their trunk to breathe, but the moderate length of their "pipe" keeps them from experiencing too much difficulty from hydrostatic pressure. We will see later why man can imitate the elephant only down to a few feet.

Other insects display incredible diving abilities, taking their air supply with them externally, trapping air bubbles in hair on their legs and bodies and bringing them down below. In some cases this breathing supply is drawn upon directly, or the bubbles are released into some underwater niche and stored for later use. There is a water spider which uses just this technique.

The common water spider (*Argyronete*) spins a web below the surface, securing a dome-shaped nest to objects on the shallow bottom of ponds. The silken web is spun underwater and is obviously filled with water, but then the spider begins the process of bringing air down from the surface. On repeated dives, the spider traps air in its slender, hooked legs on the rear half of its body. These bubbles are held secure until the spider positions itself directly underneath the web. The legs are then rubbed together, the bubble is released and rises to the top of the web, where, one by one, they displace a little bit of water each time. The spider continues laboring until the nest is completely filled with air. In the case of a female spider, when the nest is full of air, it is then time to lay eggs. For a male, he will have spun his web next to a female's and it will be time to tunnel through to her nest so that there will be a dry passage that can facilitate mating.

The spiders, with a respiratory system somewhat different than that of humans, are able to maintain an equilibrium in their underwater homes because the expired carbon dioxide is dissolved in water and pure oxygen is extracted in its stead.

The dome-shaped web of the water spider may well have been what inspired man to use a diving bell underwater. But he may also have been observing how air keeps water from an inverted pot pushed down into a tub. In any case, humans were limited in their options so long as the pump was not invented. Pumps supplementing our weak lungs made hard-hat diving possible. From then on, diving techniques made great strides.

Such a diver would suffocate unless his **breathing tube** *was linked to a pump. Human lungs can't draw air more than a few feet below the surface.*

Mosquito larvae live underwater by drawing air from the surface through a tube. Early diving designs tried to emulate this but usually failed.

Tube Tales

One of the oldest underwater breathing devices has to be the hollow reed or tube. Prehistoric hunters must have used these when they were stalking animals watering on the banks of a river or on the shore of a lake. The idea for using such a contrivance could have occurred to Stone Age man after seeing elephants standing submerged beneath the water and extending their proboscises out of the water to breathe.

Aristotle, who lived in the fourth century before Christ, explained that divers he had seen or heard of had used hollow tubes extending well out of the water. Aristotle also acknowledged man's debt to the pachyderm when he wrote, "Just as divers are somewhat provided with instruments for respiration through which they can draw air from above the water, and thus remain for a long time under the sea, so also have elephants been furnished by nature with their lengthened nostril; and wherever they have to traverse the water, they lift this above the surface and breathe through it."

Some 400 years later, the Roman Pliny reports that swimmers were using breathing tubes during long underwater excursions.

A book published in the Middle Ages, but attributed to the Roman writer Flavius

Vegetius Renatus of about 375 A.D., discusses diving devices in connection with warfare. One of the illustrations in the treatise shows a man underwater pulling down air from the surface while he is encased in a leather bag which fits tightly around the head, neck, and shoulders. The end of the flexible tube is held above water by means of a float which appears to be some sort of inflated bladder or sack. There are two different versions of the illustration, only one of which shows eyeholes in the leather contraption over the diver's head. In reality, neither could have functioned underwater.

Leonardo da Vinci, who obviously was not a diver himself, devised an apparatus which called for a man underwater to breathe through a flexible tube that was held above the surface with a disk-shaped wooden float. There is a flexible mouthpiece which is fastened behind the diver's head and opens only into the tube, which is placed at the mouth. Another design by Leonardo had a tube extending from the back of the head to the surface, in an apparent attempt to aid the diver's vision as he moved underwater.

Other designs of the sixteenth century were by Vallo and Lorini. Vallo's device was similar to Leonardo's, while Lorini's tube was a rigid pipe about a foot in diameter. All such tubes were unworkable more than a foot or so below the surface because the diver's lungs and chest muscles would be subjected to pressures so great that he would be unable to draw atmospheric air down to his lungs for more than a very few respirations. The maximum depth at which a man can inhale once or twice through a pipe is eight feet!

The grub of the drone fly, Eristalis tenax, *is called the **whip-tailed larva** (right) because of its unique underwater breathing apparatus. The grub has a telescopic tube which can be extended as much as six inches to reach air above the surface of the water.*

Siebe, Gorman & Company Ltd.

Leonardo da Vinci sketched several designs for underwater breathing tube (above). Figures A and B, with their flexible tubes, could have been used in very shallow water, but the more elaborate mechanism in figure C, complete with neckpiece and goggles, would be unworkable at the depth indicated.

Siebe, Gorman & Company Ltd.

A Sack of Air

Whether taking the hint from nature by watching other diving creatures or by relying on his own ingenuity, man began very early to devise methods of augmenting his own limited ability to remain submerged in water. One of the most practical solutions seemed to be to take air with him from the surface. This idea, if we can believe what we see in an Assyrian bas-relief, was tried as early as 900 B.C. It is questionable, however, if those bags were used actually for diving or merely as a float or life jacket: it would require very heavy weights to counteract the buoyancy of the bags and to sink the divers. There is also an illustration from the Middle Ages, attributed to about 375 A.D., depicting a man underwater breathing air from some sort of inflated sac. It may have been the bladder of an ox or some other type of compressible watertight bag. The picture shows a naked diver striding along the seabed, bag in hand and even a leafy tree growing up from the bottom, with its upper boughs breaking the surface amid placid swans.

The use of such a sack, from whatever date, would be possible only in very shallow water, where the water pressure would not collapse the bag too much as it does lungs. But the discovery of the physical laws of pressure were still very much in the future, and since the designers of diving equipment were for the most part not divers themselves, many subsequent inventions and contraptions were as weird as they were unworkable. But even if the laws of pressure were unknown, trial and error showed that flexible bags were not the best method of transporting air below the surface and that if man insisted on breathing in and out of them for some time, he would suddenly pass out, intoxicated by carbon dioxide.

Old notions die hard. Late in the seventeenth century Giovanni Alfonso Borelli sketched a design for a leather bag which could be carried on the diver's back. The diver's buoyancy was to be adjusted by means of a piston. It didn't work, but it may have inspired later inventions which eventually led to the aqualung. Another of Borelli's conceptions was a self-contained recirculating breathing apparatus in which the exhaled air was purified through cooling water.

In 1772 Sieur Freminet designed a leather diving suit that was similar to Borelli's conception of a hundred years earlier. The reservoir, which apparently was used with limited success, had a pair of tubes attached to the helmet and incorporated a spring device and bellows to pump air from the reservoir.

Flexible containers would not be the answer for a person breathing underwater, so long as scrubbers had not been invented to purify the air and oxygen could not be replaced.

*These **two designs for diving gear** were published in a 1511 edition of Vegetius's De re militari.*

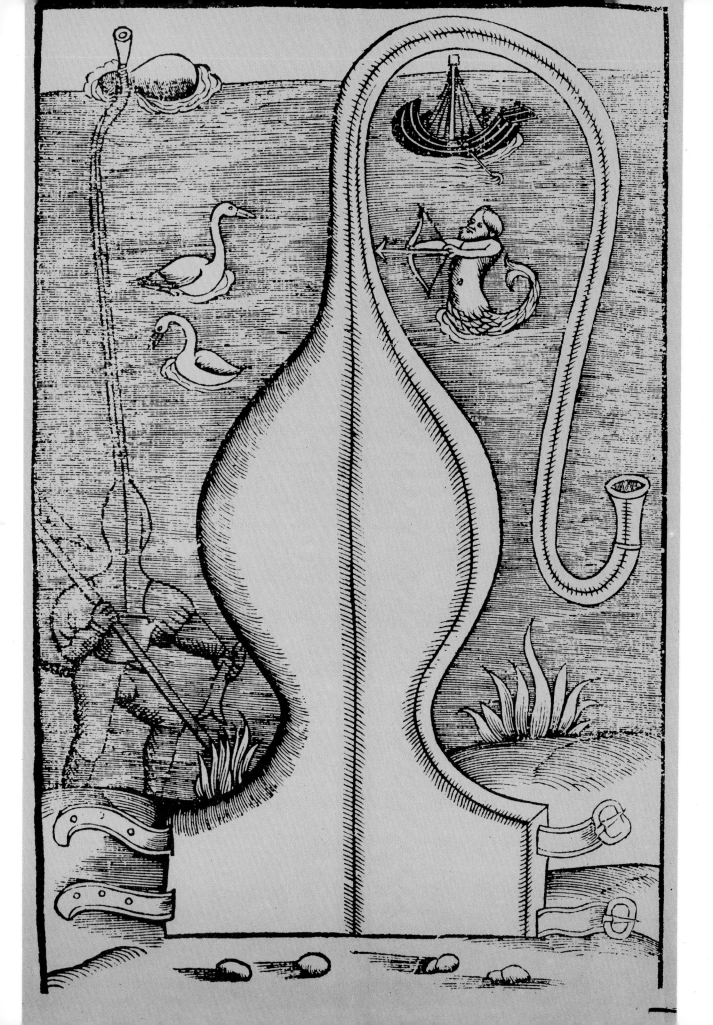

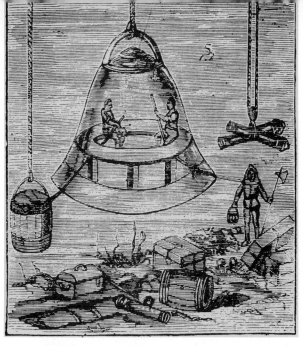

Siebe, Gorman & Company Ltd.

Edmund Halley's bell (above) *was supplied with fresh air sent from the surface in small barrels.*

Another version of **Halley's bell** *(opposite page) rests on tripod legs as divers salvage a wreck.*

Capturing Air

The rigid diving bell is a very old device, one that has been used with success, especially in shallow water. There is a problem with buoyancy as such bells would pop up to the surface if they were not secured or heavily ballasted. Heavy metal bells solve this problem, but if they are used to work on the bottom, they must be suspended to let divers swim off. The most logical answer is to hang the bell from the tender vessel with a rope. Restricted vision is another obstacle, unless the container is made of glass. The greatest difficulty is that water pressure compresses the air inside the bell as the depth becomes greater. At 33 feet down, air inside a bell will be reduced to half its original volume. The air supply inside the bell will be quickly fouled by the diver's respiration as there is no way to renew it; the worker soon has to surface, either with the bell or by ducking out under the hood and swimming free. These difficulties did not deter inventors of the Middle Ages from devising various types

of bells. Most of them were used as an air reserve to avoid constantly swimming back and forth to the surface.

Early in the sixteenth century, Guglielmo de Lorena designed a small device suspended from a ship. It was cylindrical and was equipped with shoulder braces so that the diver could steady the apparatus while keeping his hands free. There was a glass window for observation purposes. Lorena's device was actually used in a search for sunken ships in Lake Nemi south of Rome.

In 1616 Franz Kessler used the bell concept for a device which covered the diver down to the ankles. Kessler's design was so much like a bell that it had a large clapper inside that was used as ballast.

At about the same time as Kessler's design was made, Francis Bacon described another device that was in current use. "It was a hollow vessel made of metal," he said, "which was let down level to the surface of the water and thus carried with it to the bottom of the sea the whole of the air which it contained." The device stood on three feet like a tripod. The dome of the contraption stood about five feet above the floor so that, as Bacon explained, "the diver when he was no longer able to contain his breath, could put his head in the vessel, and having filled his lungs again, return to his work."

The efficiency of the elementary bell concept was best demonstrated in 1687 when William Phips used them to recover the fabulous treasure of the sunken Spanish galleon *Nuestra Señora de la Concepción* on the Silver Banks in 35 feet of depth.

The most successful bells were built in 1690 by Edmund Halley, the astronomer who discovered the cyclical nature of the comet that bears his name. The air inside the wooden, heavily ballasted bell was renewed by smaller barrels sent down from the surface.

65

*In 1772 **Sieur Freminet** invented a **diving apparatus** consisting of a leather dress and copper helmet. Air was supplied by a reservoir worn on the back and connected to the helmet by two tubes.*

Pressing the Search

Man's early efforts to reenter the sea were drastically limited in time and depth by his inability to force large quantities of air to supply the diver in a safe, practical routine.

In 1660 Robert Boyle had clearly defined the physical laws governing gases under pressure. As a consequence, in 1689 the inventor Denis Papin, a pioneer in the development of the steam engine, had the idea of a crude air pump to supply a diving bell continuously. The air would thus be forced down the rim of the bell whatever its depth, instead of rising inside it at a level varying with depth, which reduced its buoyancy and the space available for divers. The constant flow of fresh air would greatly extend the duration of the stay. It was a revolutionary concept in diving techniques.

It was only about one century later that John Smeaton constructed a bell that was the first practical realization of Papin's suggestion. It was not totally submerged, and air was pumped through its roof. The news spread quickly, and soon in France, England, and Germany there was an explosion of inventions and a quantity of actual constructions and tests of still imperfect but usable devices. Sieur Freminet in the Seine River, K. H. Klingert in the Oder, and William Forder in England claimed successful dives with equipment differing greatly, but all basically including a rigid helmet fitted with windows and attached to a leather suit; they were fed with air at a pressure higher than atmospheric pressure by a pipe connected with bellows, another pipe being used to evacuate fouled air.

A more imaginative device for using compressed air from an underwater reservoir was devised in 1809 by Frederic de Dreiberg. The diver was supposed to carry a watertight container on his back and wear a crown on his head. The crown was attached to the air container through a series of rods which in turn worked a bellows on the reservoir, compressing the air and forcing it through a breathing tube into the diver's mouth. It certainly could not function!

An underwater artist leisurely works at his easel *(right). He wears a suit that is a modern version of Augustus Siebe's original, built in the nineteenth century. The air hose supplying air from the surface is clearly visible.*

Siebe's first **open diving helmet** *(below). Air was pumped into it from the surface and excess air escaped from beneath the jacket at the waist.*

Siebe, Gorman & Company Ltd.

Ten years after Dreiberg unveiled his invention, which he called Le Triton, Augustus Siebe displayed his open diving dress which consisted of a hard helmet and leather jacket. Air was pressure-pumped into the helmet from above, and the exhaled gas and excess air escaped from beneath the jacket at the diver's waist. This very simple design allowed the diver to keep his head above water in what could be described as an individualized diving bell. The apparatus was practical to a certain extent and was much safer than all the former devices and could be used at any depth so long as the air pump was powerful enough. It was used in a number of successful salvage operations.

Hard Hats: The Mini Bells

For a little more than a century, the standard equipment of a diver was basically a miniature diving bell (called a hard hat) connected to an air pump at the surface by a hose. Heavily weighted, the diver remained in an almost vertical position and had to walk clumsily on the bottom. Modifications and improvements in hard hats and diving suits came quickly in the first half of the nineteenth century. In 1834 an American, L. Norcross, invented a diving suit that enclosed the diver entirely with an air outlet valve in the shape of an inverted siphon on the top of the helmet to eliminate foul air. A year later, J. R. Campbell suggested using

a helmet made of glass, in the candid hope of improving vision. Campbell's suit was rigid to the waist and flexible below; the exhaust pipe had its outlet at belt level.

In 1837, Augustus Siebe presented his closed diving suit, which really was an alteration of his earlier open diving dress. The helmet and forced-air pump were essentially the same as in Siebe's earlier model, except that the new helmet was fitted with an outlet valve, but now the diver didn't have to get wet. Siebe's closed diving suit became the basic model for all other diving suits.

*Augustus Siebe's **closed diving suit,** built in 1837, became the model from which later suits were designed. The pump provided the diver with air.*

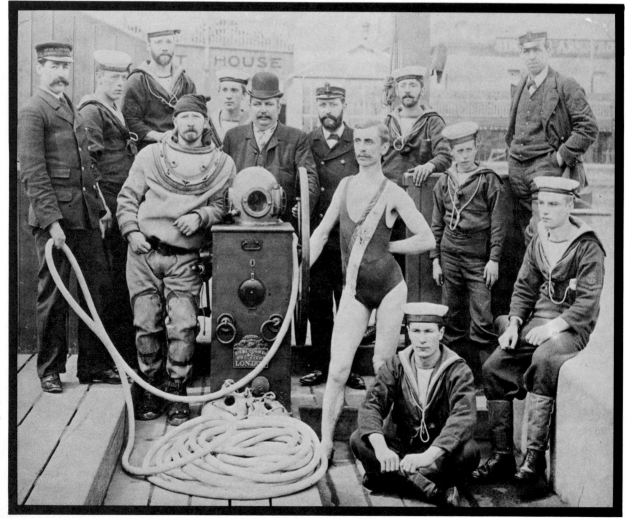

Siebe, Gorman & Company Ltd.

on offshore oil-drilling operations, in salvage work, and in harbor maintenance. The colony of Greek sponge divers in Tarpon Springs, Florida, long ago converted from the naked diving of their ancestors to hard hats and closed suits, soon after their Mediterranean colleagues had done so.

The modern outfit, which weighs nearly 200 pounds, consists of a helmet, breastplate, weightbelt, flexible watertight dress, and weighted shoes. Those boots are made of either brass, which is good for walking around in mud, or lead, which provides better traction on solid surfaces like ship's hulls. The brass shoes weigh 20 pounds a pair; lead, 35 pounds a pair.

The umbilical connection to the surface includes the air hose, a lifeline, a telephone line for intercom between the diver and the tender. The air hose and the lifeline remain separated for safety reasons; the telephone line is either attached to the air hose or included in specially braided lifeline ropes. The depth of the diver is known by the crew above by reading on a gauge the pressure of the air that is pumped down.

Hard-hat diver and free diver (below) move side by side. The development of the aqualung freed divers from cumbersome hard-hat gear and gave them more mobility under the sea.

A navy diver (above) has his suit checked before he submerges. An air hose, lifeline, and telephone line connect the diver with the surface. The modern diving suit weighs about 200 pounds.

There have been alterations in Siebe's design, but none of them have been so major as to change the basic pattern. Hard-hat divers in flexible suits—as opposed to the armored articulated diving dress—work at impressive depths: to 200 and even 300 feet with air, and to 600 feet and more using helium-oxygen mixtures. They still are used

Chapter VI. Freedom in the Sea

Attempting to become free as fish in the sea, we had to forget our fears and to adapt our minds to an environment that had little to do with ours. Obviously we had first to get rid of those hoses and lines that turned us into captive animals kept on leashes. Then we had to abandon walking on the bottom because of two reasons: first, we wanted freedom in three dimensions, and, second, the vertical walking posture was offering a maximum resistance to propulsion and to currents in such a thick element as water. Therefore we had to develop a practical self-contained breathing apparatus; and we had to

---

"Air hoses and lifelines were symbols of our fears and misunderstanding of the sea."

---

move by swimming in a longitudinal position and to be neutrally buoyant.

To improve the realm in which we would exercise that freedom and the length of our excursions, we had to acquire a better knowledge of diving physiology.

Air hoses and lifelines, symbols of our fears and of our misunderstanding, had in fact often proven to be deathlines. Many inventors had dreamed of getting rid of them: projects from Klingert in 1797, Fullerton in 1805, and James in 1825 were self-contained with a built-in air supply.

In 1869, Jules Verne's *Twenty Thousand Leagues under the Sea* foresaw at least a two-dimensional freedom of men reentering the sea. Captain Nemo, skipper of the submarine *Nautilus,* remarks to Professor Aronnax, "You know as well as I do, Professor, that man can live under water providing he carries a sufficient supply of breathable air."

But already in Verne's time, practical steps were being made toward freedom. Four years before the publication of Verne's classic, Benoit Rouquayrol and Auguste Denayrouze unveiled a device which allowed a diver to store a small amount of compressed air on his back, disconnect his air hose to the surface, and walk free on the floor of the ocean. The freedom was short-lived and the apparatus primitive, but those first steps had been taken in 1865. The key to the Rouquayrol-Denayrouze device was a regulator which helped control the flow of air from the underwater reservoir to the diver's mouth.

There were other steps, and missteps, along the way. One was the oxygen-rebreathing apparatus, invented by Henry Fleuss in 1878. These devices feed only oxygen to the diver, and his exhalations are filtered through a chemical agent to purge them of carbon dioxide. We will see later why they should be used only in extremely shallow water.

A more practical idea is to provide the diver with a self-contained fully automatic underwater air-breathing apparatus. Not that breathing compressed air doesn't present limitations, which had to be learned the hard way; but even within such restrictions, a tremendous step forward was made and turned diving into a popular sport. To mitigate the anatomical inadequacy of the human body to swim efficiently and without effort, foot fins and mechanical propulsion aids have been produced. But we have a long way to go if we ever want to match our gentle cousin the porpoise in his familiar element.

Jules Verne dreamed of man being able to move about under the sea. **Stinging jellyfish** *symbolize the dangers that await man under the waves.*

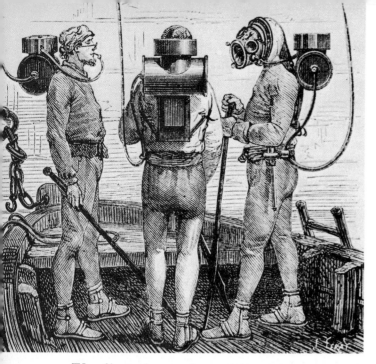

The first air regulator (above) *was developed in 1865 by Rouquayrol and Denayrouze. The air reservoir was kept full by a pump above.*

Air Regulations

Rather than gulping air and plunging into the water for a two-minute stay, divers in the nineteenth century put up with the bulky helmets which utilized air pumped down from above. The flow was more or less continuous, as steady as the manual pump operators could maintain. It wasn't until 1865 that Benoit Rouquayrol and Auguste Denayrouze developed a regulator called an "aerophore," which helped adjust the supply of breathing gas to the diver. The regulator, which was used first with a mouthpiece and no suit, and then in conjunction with a tight-fitting rubberized suit, was connected to a reservoir carried on the diver's back. This reservoir, in turn, was continually being filled with pumped air from above, at a pressure of 600 psi. The regulator helped control the flow of air between the small reservoir and the diver's mouth. The Rouquayrol-Denayrouze apparatus made breathing easier; the diver could disconnect his air pipe for a while, which enabled him, for example, to enter a shipwreck safely.

The unrestricted freedom of movement on the floor of the ocean had always been the dream of divers. With air hoses tied to a mother ship, they could walk in a circle only as large as the length of line would permit. Strong currents would act on the lines as well as on the diver to drag him away from his work. Even the early air reservoir ideas such as Klingert's tank and Halley's bell allowed only slightly more freedom and they had the disadvantage of being awkward to use and expensive to maintain.

In developing their regulator, Rouquayrol and Denayrouze were on the right track. The

most important aspect of this regulator was not that it was equipped with a detachable feed pipe, but that for the first time, self-carried air was delivered to the diver at the same pressure that the ambient water was exerting on his body. The device also provided the diver with the exact quantity of air he needed, for whatever task he was performing, which was a substantial economy.

A valve in the regulator functioned so that when the air pressure inside the reservoir was greater than the water pressure—such as when the diver was ascending—a membrane automatically sealed off the passage.

On the other hand, when the water pressure increased sharply, the valve was forced open and provided additional air to the diver.

The Rouquayrol-Denayrouze apparatus still required a crew of tenders on a ship at the surface, and the air lines were still a necessity in order to fill the small underwater air reservoir. But their regulator set the stage for important developments in freeing divers from overhead attachments.

*Modern divers depend upon a Cousteau-Gagnon **regulator,** designed in 1943, to provide them with air as needed.*

Oxygen Breathing

As long as high-pressure compressors had not been developed, inventors could not even consider taking down appreciable supplies of air. As early as 1842, a Frenchman named Sandala suggested a self-contained breathing apparatus that recirculated the air and filtered out the foul products of respiration.

The only gas really necessary to human life is oxygen. In breathing atmospheric air, man finds carbon dioxide harmful in large concentrations, and nitrogen neutral. So if anyone wanted to take Sandala's suggestion seriously, the only important components in a closed-circuit breathing system would be a reserve of oxygen and some kind of agent to absorb the carbon dioxide.

The first workable device of this kind, called an oxygen rebreather, was made in 1878 by Henry A. Fleuss. The apparatus was not really made for diving but for use in toxic

*This **early escape apparatus** (above) is being replaced by a system (left) wherein escapees wear nose clips, life jacket, and immersion suit.*

atmospheres such as coal mines where poisonous gases and noxious fumes had accumulated. During World War I, Fleuss-type breathing systems were used by the Allied military for mining or tunneling. Robert H. Davis, the diving authority and historian, adapted the rebreather for use by fliers.

The basic design was a face mask, a flexible breathing bag, and a cylinder of oxygen under 30 atmospheres of pressure, that is, about 450 pounds per square inch. The key part of the system, a container of carbon dioxide absorbent, was carried on the back, behind the breathing bag. Today the absorbent cartridge is included in the bag, the oxygen cylinder under the bag, and the entire apparatus is strapped on the chest.

A similar apparatus was adapted by Davis for emergency use in submarines. The Davis submerged escape apparatus also includes an emergency oxygen capsule, a flexible air hose, a mouthpiece, and a nose clip. The flexible bag allows the oxygen to be breathed at the pressure of the water and the breathing bag can be used as a flotation device.

Human lungs need only about a quart of oxygen per minute, regardless of depth; thus the oxygen rebreathing systems are not bulky or cumbersome, one of the attractions they hold for military divers. In World War II the frogmen used oxygen rebreathers and probably pushed them to their limits, learning that pure oxygen under pressure causes dangerous convulsions and that prolonged oxygen dives were only safe near the surface and no deeper than 23 feet.

Within this limited range, since the system is closed-circuit with little or no gas escaping in the water, its operation is unnoticed (which is appreciated by the military).

The silence of oxygen rebreathers helps scientists to observe the behavior of marine creatures without disturbing them.

Oxygen rebreathers are worn in front (left). These compact units (below) allow no gas to escape and are used by military men and scientists.

Siebe, Gorman & Company Ltd.

Getting unto Thyself

Even while divers and scientists were still experimenting with manual pumps for sending air down below, W. H. James was thinking in terms of a self-contained breathing apparatus unconnected with anything at the surface. James's conception of the device provided a reservoir of compressed air in a metal container the diver wore around his waist. The air supply was small and the diving time was necessarily limited.

The problem with James's concept was that his thinking was very far ahead of the technology of his day. The Rouquayrol-Denayrouze regulator was still 40 years in the future, while the idea of putting air under pressure in steel containers—even for industrial use—would have to wait for another century. And the reality of using a bottle of compressed air as a breathing gas underwater was almost beyond comprehension.

It wasn't until 1933 that Yves Le Prieur, a commandant in the French navy, devised a system which incorporated high-pressure compressed air in a container carried over the chest. The air flowed constantly through a regulator and a breathing pipe into a full face mask. A hand-operated valve allowed

the diver to regulate the flow in order to cut down on waste. The steady stream of air restricted the length of submergence, but Le Prieur's apparatus allowed men to stay very safely 25 feet under water for 20 minutes, and 40 feet below for 10 minutes.

Ten years later, another Frenchman, Georges Comheines, tested a semiautomatic regulator attached to a compressed-air container. It was a modified version of an air-breathing apparatus used by firemen in toxic atmospheres. Unfortunately, Comheines died during one of his first dives. By that time, Emile Gagnan and I were already working on our

*William James invented the **first self-contained diving dress** (above left) in 1825. Air was obtained from the iron cylinder worn around the waist like a belt, and the helmet was copper or leather.*

A self-contained suit *designed by Louis and Auguste Boutan (below) featured a hand-operated valve that allowed regulation of air supply. Le Prieur incorporated this into his design in 1933.*

*The **demand regulator** (above) is part of the aqualung (right) and allows a diver the freedom to inhale and exhale normally.*

fully automatic aqualung which supplied air upon demand to the diver at exactly the appropriate pressure. We were on the threshold of true freedom beneath the sea.

The aqualung is based on the principle of open-circuit breathing apparatus, which allowed the exhaled air to escape into the water; this is a waste of oxygen but is only the price of simplicity and safety. The closed-circuit oxygen rebreathing devices on the contrary allow the diver to use up all the oxygen available. I had started my experiments with such devices and had personally had two serious oxygen-poisoning accidents in only 50 feet of depth. Our open-circuit aqualung proved to be the most popular passport to the silent world of the sea.

Ask and You Shall Receive

To achieve complete freedom under the sea within the limits of the complex physiological laws that will be reviewed later in this book, the ideal diving apparatus has to provide a substantial reserve of compressed air. It must supply this air at exactly the same pressure which water exerts on the chest; deliver automatically all the air needed but nothing more; and maintain a comfortable ease of breathing for the diver in any position. And the apparatus has to be light enough to allow the user to remain neutrally buoyant as a cartesian diver.

The diving bell, the open-bottomed helmet, Siebe's hard-hat suit, the Roquayrol-Denayrouze and Le Prieur gear had paved the way without fulfilling all the above-mentioned requirements. Industrial technology was now offering high-pressure compressors and lightweight air cylinders; de Corlieu was introducing the swim fins. All that was needed was a practical, well-adapted "demand valve" to control automatically the flow of air the diver breathed.

During World War II, in occupied France, engineer Emile Gagnan designed such a valve to feed the motors of cars with cooking

*Six men prepare for a dive using Emile Gagnan's demand valve as a **breathing apparatus**.*

gas! With a few modifications, it became the essential component of the aqualung—or independent compressed-air breathing apparatus—which revolutionized diving.

In this system, the compressed-air cylinders are carried on the back. The air passes from the container through a control valve which then brings the pressure down to about 100 psi above the surrounding pressure. Then the air passes through the demand valve operated by a membrane which is subjected, on the outside, to ambient water pressure. The air on the inside of the membrane is soon equalized with the hydrostatic pressure. Each inhalation applies a small depression on the large surface of the membrane, which acts as a force multiplier and pushes the valve open: air is fed to the diver. As he exhales, the membrane falls again, a spring closes the valve, and the air flow stops. The exhaled foul air escapes freely into the water through a one-way exhaust valve and never mixes with the air to be inhaled.

But in the sea, things are not quite that simple. If the membrane of the demand valve is not fastened quite close to the center of

TANK
VALVE

AIR
TANK

F

E

EXHAUST HOSE

D   C

A

B

INTAKE HOSE

H

G

gravity of the lungs, the ease of breathing will vary greatly with the posture of the diver. When the demand valve is below the lungs, the diver will receive a free flow of air; but if the membrane is above the lungs, the diver will have to draw his air painfully. Both situations are very uncomfortable.

The location of the outlet nonreturn valve at the extremity of the exhalation pipe is also very critical. If it were above the lungs, the regulator would be constantly activated, and the air would be constantly flowing and wasted. The outlet has to be fastened within an inch or two of the center of the membrane. A simple warning system alerts the diver when the air supply is low and assures him

*How the **demand valve regulator** works. Inhalation reduces pressure in chamber (A) above membrane (B), causing it to rise. A lever (C) is activated by this membrane, allowing air to flow through the low pressure valve (D). A reduction in pressure is caused within chamber (E), opening the high pressure regulation valve (F) and allowing air to flow to the diver. Expired air flows through a one-way valve (G) into chamber (H), which is open to the water through the holes indicated (dotted lines).*

of a reserve sufficient for a return to the surface. The aqualung diver cannot breathe anything but pure air. In many other systems, carbon dioxide from exhalation can get mixed in the helmet or the mask with the inflowing air, or it might not be totally absorbed in an oxygen-rebreathing apparatus.

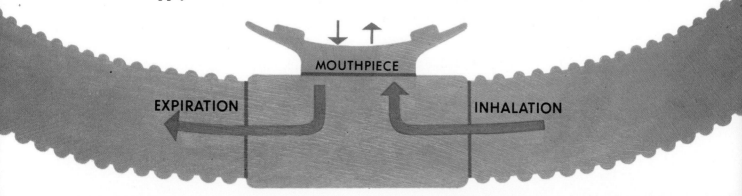

MOUTHPIECE

EXPIRATION

INHALATION

Don't Hold Your Breath

The old law of gravity states that whatever goes up must come down. In diving, this axiom may be worded: whoever goes down must come up—with caution.

When a free diver leaves the surface without a breathing apparatus, he has a given volume of air in his lungs which gets smaller and smaller the deeper and deeper he goes because of the change in hydrostatic pressure. This process is reversed when the diver ascends. The volume of air in the lungs expands as the water pressure around him decreases. This process is entirely natural and is ordinarily safe.

But the diver who breathes compressed-air underwater introduces an additional quantity of air into his lungs, which will expand as he ascends. If the diver stops breathing, his glottis acts as a tight valve and this expansion of the trapped air in his lungs can distend them beyond their maximum capacity, causing lesions in the lungs and forcing gas under pressure into the bloodstream, resulting in embolisms. When air embolisms occur, high-pressure air bubbles introduced in the bloodstream expand quickly and can block the circulatory flow to the brain and to major organs. They can cause serious damage and quick death. If the brain or nerves are affected, pain and paralysis may occur. At their mildest, embolisms cause pulmonary irritation and coughing. Air embolisms are terrible accidents that can occur very easily if a diver holds his breath during his ascent, as when he fears running out of air before reaching the surface.

*The beauty of the undersea world can be awe-inspiring, but a good diver never forgets his physical limitations. A hasty **ascent from the depths** can result in air embolism: bubbles forced into the bloodstream that can block circulation.*

A common diver-training exercise requires the novice to abandon his equipment on the bottom and return to the surface. This is a very dangerous practice and should be prohibited because there is a tendency for beginners to hold their breath. Only trained divers, totally at ease in the water, should attempt this exercise and then only under very close supervision.

Fatal embolisms can occur in such manner in dives of modest depths, from as little as eight feet down. After deeper dives, the problem is usually faced within the last few feet of ascent, when the diver experiences relative variations of pressure very rapidly. Often a stricken diver is conscious upon arriving at the surface, but loses consciousness soon thereafter if the accident is serious. This reaction is often mistaken for surface drowning, which requires an entirely different treatment and first aid. Embolism victims must be immediately recompressed. This allows the gas bubbles in the blood to be reduced in size as the pressure on the entire body is increased; the circulation of blood is reestablished and recovery is spectacular. This can be followed by a relatively brief decompression and by usual treatment for pulmonary lesions.

Many victims of air embolism are neglected or improperly treated because of the assumption that their dives were not deep enough or long enough to qualify as decompression accidents. The whole problem can be alleviated by ascending slowly and exhaling during the rise to the surface. Air embolism is a direct result of holding the breath during ascent; it cannot occur when a diver breathes normally while ascending. The rule of thumb is never to rise to the surface faster than the stream of small bubbles do from the aqualung exhaust valve. Humming or singing during ascent, which keeps the glottis open, is another good safety rule.

Nitrogen and the Bends

Nitrogen is considered a neutral gas in the air since it does not perform any essential function in the life process and does not react with the human body. At sea level, the body is penetrated and saturated by the surrounding air. Approximately one quart of nitrogen is dissolved in the blood and tissues at normal atmospheric pressure.

When a diver descends, the amount of nitrogen dissolved in his body increases proportionally to the increase in pressure. And when a diver breathing compressed air for some time and at some depth rises, enough nitrogen has been dissolved in his system to form bubbles as the pressure decreases—in

*In the process of **stage decompression**, a diver checks his watch to see if it is time to continue his ascent. He also wears on his arm a Fathometer to record depth and a compass for orientation.*

much the same way as bubbles form and stream to the surface when a bottle of soda water or champagne is uncapped. These beverages are bottled under pressure and are decompressed when opened.

Such bubbles present special hazards when a diver has been submerged for a relatively long period or has dived to relatively great depths. When the behavior of those bubbles results in compressed-air illness, it is usually referred to as the "bends."

During a dive when additional quantities of nitrogen penetrate the diver's body, it will take about 6 to 12 hours before a saturation point is reached corresponding to the new pressure level. Since fatty tissues can absorb more nitrogen, it takes fat people longer to become saturated. It also takes them longer to decompress, if they should stay under long enough to become nearly saturated. The normal quart-load of nitrogen in the bloodstream increases with pressure—at 33 feet where the pressure is at two atmospheres the body when saturated will have about two quarts of dissolved nitrogen. If the diver returns rapidly to the surface from two atmospheres of pressure, his body tissues and fluids are then saturated with twice the amount of nitrogen, and this excess can be eliminated quite adequately through the circulatory system and lungs until the nitrogen level returns to normal. If the diver had been subjected to pressures greater than two atmospheres—that is, a depth greater than 33 feet—nitrogen is dissolved in such a concentration that it cannot be readily eliminated by the body: the result could be the bends. Nitrogen is the villain, although an excess of carbon dioxide in the system aggravates the situation.

Such accidents as the bends can be avoided only by the diver making a very controlled ascent, slow enough to eliminate excess dis-

solved gas gradually so that oversaturation does not reach the critical point where dangerous bubbles form.

The best-known method of ascent, called stage decompression, was proposed by J. S. Haldane in 1906. Stage decompression requires the diver to stop rising every ten feet for specified time periods, which are determined by the depth and the duration of his dive. The length of time of each of these stops, which increases near the surface, are listed in decompression tables.

Four divers decompress on the anchor line after a 200-foot dive. A timed ascent allows the body to dissipate dissolved gases gradually, thus avoiding bubble formation in the bloodstream.

During the frequently repeated descents of native pearl divers to 100 or 120 feet, each time a little more of the nitrogen contained in the lungs dissolves in their blood. After a number of dives, this quantity may become sufficient to cause a decompression accident upon return to the surface. The Tuamotu islanders call this sickness "tarawana."

Diving Physiology

Two men of science who have had the most impact on modern diving are Frenchman Paul Bert and Scotsman John S. Haldane. In 1878 Bert published his classic, *La Pression barometrique,* in which he noted, "Pressure acts on living beings not as a direct agent, but as a chemical agent changing the proportions of oxygen contained in the blood, and causing asphyxia when there is not enough of it, or toxic symptoms when there is too much."

It was Bert who early discovered the effect of breathing nitrogen under pressure—it increases in the bloodstream since it does not pass through the body as a normal waste product at exhalation. Bert realized that as the nitrogen bubbles came out of solution they could block blood circulation and cause the bends. The impact and importance of Bert's work was such that his thousand-page treatise was still being used as a textbook 65 years after it was first published.

It was Bert who suggested that the only way to avoid the bends was to ascend slowly, so that the body could gradually decompress— rid itself of excess nitrogen. Bert died at the age of 53, before he could pursue his studies to the fullest. There was a lapse of more than

Paul Bert was a French scientist who declared that water pressure influenced the amount of dissolved nitrogen in the blood. He suggested ascending slowly to avoid decompression accidents.

John Scott Haldane studied nitrogen saturation and decompression and devised tables telling divers how long to wait at a specific depth to let nitrogen come out of solution, thus avoiding the bends.

ten years before Haldane was able to do some practical experimentation on nitrogen saturation and decompression.

Haldane, a doctor of medicine, became interested in the problem of breathing in toxic atmospheres following a coal mine disaster. His early work was with carbon monoxide, and with carbon dioxide poisoning, where he found that normal breathing depends solely on the pressure of carbon dioxide in the respiratory center. Eventually, Haldane became interested in divers who were constantly being subjected to varied and artificial atmospheres. After working with Royal Navy divers, Haldane established decompression tables that specified waiting stages for ascending divers, stops they must make to allow nitrogen to come out of solution in the body and pass through the system. Haldane devised his tables on a theoretical basis and then altered them as needed after working with the divers. His tables were graded for each 12 feet—two fathoms—of depth,

Physiologists intently monitor the dials and gauges of this pressure chamber. In research on the effect that gases might have on divers, animal subjects are exposed to mixtures of breathing gases of varying concentrations and pressures.

but the waiting stages were graded at every 10 feet. Haldane's original tables went down only as far as 204 feet, which was about the known limit for hard-hat divers at the turn of the century. The diver could rise directly to 80 feet before stopping, but then had to stop every 10 feet for increasing periods of time. The longest stops were near the surface, at the 20- and 10-foot marks. An ascent from 200 feet, requiring eight stops, could take as long as 238 minutes.

Haldane's tables have been altered and expanded as greater depths have been reached. But his work in the first decade of this century, and Bert's efforts before him, helped form a foundation for today's diving art.

A group of worried divers anxiously await news about their companion undergoing decompression. **Decompression chambers** *onboard ship have saved the lives of many undersea explorers.*

Decompression Accidents

Decompression tables giving duration of stops for each stage of decompression are now maintained by most navies of the world. They are established for divers using compressed air and doing moderate work. Dives involving great muscle activity increase the rate at which nitrogen is absorbed by the body, and divers would be advised to decompress as though a dive had lasted longer than it actually had.

There are several kinds of accidents which can occur by not following decompression tables and procedures. The most serious involve a release of bubbles into the blood as the diver ascends, bringing on a blockage of pulmonary circulation, which can be fatal within a short time. This release of bubbles is similar to the phenomenon of gas bubbles appearing in a bottle of carbonated soda when it is opened and exposed to normal air pressure. Other severe damage can happen as a result of oxygen being cut off from the brain by partially interrupted circulation, such as hemiplegia, paraplegia, or other extended paralysis.

Other, less severe effects of decompression accidents can be suffered as results of long dives without adequate decompression: localized pain in the joints which often forces the diver to double up—the bends. A minor affliction called the "fleas" is an itching sensation on the skin.

The treatment for all these accidents, without exception, is immediate recompression. This return to a higher surrounding pressure reduces the size of the bubbles, reestablishes circulation, and restores the function of nervous tissues. If the starvation of these tissues was not too serious and if recompression was accomplished in time, the accident's effects disappear entirely within several hours. If these conditions are not met, however, there will be irreversible lesions and permanent damage. And even divers who have been involved in only minor accidents may develop, with time, necrosis of the bones, especially joints at the shoulder and hips.

Recompression of a diver who has surfaced too quickly can take place in the water, but this can be difficult and even dangerous—especially if he is in great pain or otherwise injured; in any case, recompression in water exposes the victim to additional stresses from cold, and to hazards from changes in weather conditions: it is better than no recompression at all, but a decompression chamber is superior.

The quicker an affected diver is recompressed, the more spectacular his recovery. But experience shows that a certain degree of improvement is often obtained even many hours after the bends were noticed. Accordingly, one should never state that it is too late for a recompression treatment.

*While exploring the Red Sea in 1955, **one of Calypso's divers** felt symptoms of the bends and had to be recompressed. The quicker a diver is recompressed, the more spectacular his recovery.*

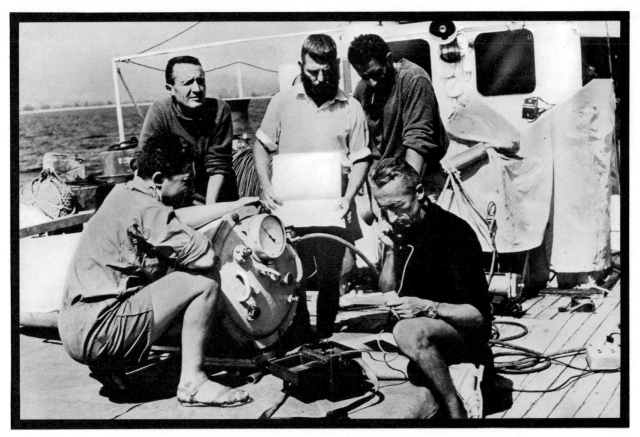

The Heat Machine

One of water's most remarkable properties is its ability to absorb heat, as you can see for yourself on any hot day when people flock to the beach to cool off in the sea. On the other hand, the human body is a heat machine, giving off heat produced by chemical reactions inside of it. In the atmosphere, the body is cooled by convection and evaporation—water passing through pores—and exhalation of the breath, which includes water vapor. When a body is immersed in water, however, it loses heat by conduction much more rapidly than on land, because water absorbs much more heat than air. If a diver or swimmer is not sufficiently insulated, his body loses heat faster than he produces it. Or, more simply put, he will be cold.

Water's ability to absorb heat has been known by experience since ancient times,

but most of the early diving-dress designs were more concerned with the problems of pressure and waterproofing than with loss of body heat. Part of the reason is that primitive divers didn't stay in the water very long. Pearl and sponge divers wear only a loincloth and generally work in waters that are warm at the surface. Their stay below is usually less than 90 seconds, and though they often repeat such dips, they take the time to warm up in the sun.

Many long-distance swimmers felt that greasing their bodies would help cut down on heat loss, but the grease washes off quickly and leaves only a thin, oily film which slightly *increases* the rate of heat loss from the body by radiation.

Retention of heat is important to divers, since cold tends to slow actions, impairs tactile sensation, and requires the body to burn more oxygen as it tries to warm itself. In freezing polar water, a swimming castaway can survive less than 15 minutes.

As a rule of thumb, a protective suit is necessary for diving in water below 64° F. The rate at which the body loses heat is ordinarily determined by the temperature of the surrounding water and the thermal resistance of the body and the diver's clothing. Wool affords excellent protection against body heat loss, but only if it stays dry. Suits of plastic neoprene foam maintain their insulating qualities, even when wet. There are "wet suits," which allow some water to seep in, and "dry suits," which have tightly sealed openings at the wrists, face, and neck. The efficiency of protection in all cases is due to the air trapped either inside the suit or as tiny bubbles within the neoprene foam. Modern dry suits give excellent protection for more than an hour in the coldest waters of the world. But if air is replaced by helium, as happens when oceanauts dive from deep chambers or stations filled with heliox, they do not protect anymore, because helium conducts heat seven times better than does air. The consequences can be very serious and even fatal (Sealab III, *Sea-Link,* Keller's California record dive, etc.). For such special test or individual dives, there are diving suits heated either electrically or by hot water pumped through them.

*Captain Cousteau dons a **foam rubber wet suit** (above). The diver below wears similar gear. Heat insulation is maintained by tiny air bubbles in the foam and water trapped inside the suit.*

***Frederic Dumas and Captain Cousteau** (left) try to warm themselves after cold-water dives. The human body loses heat more rapidly in water than in air. A protective suit is needed in water below 64° F., for any extended period.*

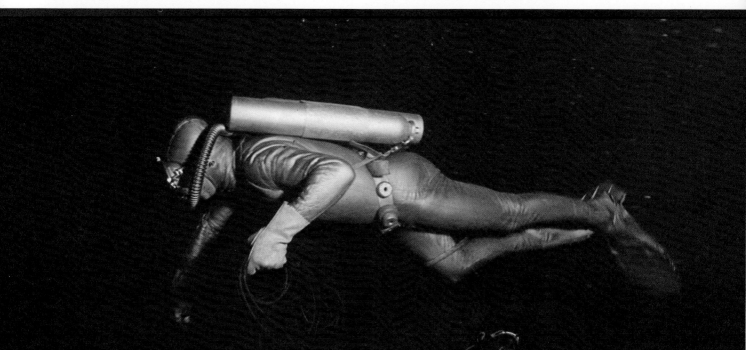

Feeling High Deep Down

Divers who surpass the 100-foot depth while breathing compressed air may find that their efficiency begins to diminish and may even begin to experience strange sensations. At 150 feet down, the faculties are obviously impaired and below 200 feet there is a feeling of being drunk. Below this level the intoxicating feeling of nitrogen narcosis may become violent enough to result in an epileptic crisis followed by a loss of consciousness. The seizure and loss of consciousness may be as much due to the high partial pressure of oxygen in compressed air as it is to nitrogen narcosis.

In its early stages and when its effects are first being felt, nitrogen narcosis results in a subtle loss of judgment, a feeling of well-being, and a reduction in the ability to concentrate—like drinking a little too much alcohol or smoking marijuana. As a result, we first called this effect "rapture of the deep." I am susceptible to it, and yet I fear it like doom because it destroys the instinct of self-preservation. There are no aftereffects from nitrogen narcosis, however; to escape its effects, one simply ascends 30 or 40 feet; one's head clears instantaneously and one is no worse for the wear.

As with alcoholic inebriation, there is a variation in symptoms of nitrogen narcosis from one individual to another, and even in the same individual from dive to dive. But just as the senses play tricks on the inebriated individual, so does rapture of the deep affect the human body, with double images, exaggeration of some senses and attenuation of others, impression of levitation or crushing, waves of euphoria or terror, and adverse effects on mental judgment and physical coordination. The wide array of symptoms shows how difficult it is to accurately measure the effects of nitrogen narcosis. It is easy to confuse it with other phenomena, and a well-trained and strong-willed diver is often able to compensate for his impaired ability by concentrating intensely on the tasks that he is supposed to perform. Those who do not realize or do not admit such loss of acuity are probably in a more serious state and should be supervised.

There are several theories available to explain the cause, but none of them is totally satisfactory. It is thought by some that the increased density and viscosity of gases at great depths reduces the efficiency of lung ventilation and results in the retention of carbon dioxide which would cause "rapture." But this is an oversimplification and contradicts the results of experiments that show that while carbon dioxide can aggravate the "rapture" symptoms (as actually happens when a diver exerts himself), carbon dioxide itself is not sufficient to induce them. Narcosis is not found in a fatigued diver breathing a mixture of helium or neon and oxygen, but if nitrogen is replaced by argon or neon, it will be more severe than with air.

Nitrogen narcosis, or "rapture of the deep," is not always obvious. We feel its effects at depths of 100 feet, and they become more pronounced as a diver descends deeper. Generally there is a feeling of euphoria and a loss of instinct for self-preservation.

Underwater Aids

Entering the sea, a naked human being finds himself blind, vulnerable to cold, poorly streamlined, and lacking adequate propulsive organs. As far as diving ability, the figures given on pages 56–57 prove that if men spent their lives in the sea from the point of birth on, they would probably be able to dive almost as far down and as long as a porpoise of comparable size.

We saw in the volume *Window in the Sea* that face masks can correct the diver's vision and that cameras supplement it. Adequate suits reviewed on pages 88–89 are able to protect him against cold and improve his streamlining. But his arms and legs do not lend themselves to any efficient propulsion method. In general, there were few improvements in swimming aids before the sixteenth century, when one of the foremost minds of the times devoted his energy and talents to the problem. Leonardo da

Vinci's notebooks included designs for various types of snorkels and swim fins. The latter were for the hands rather than the feet and resembled webbed claws. There is no report that these hand fins were ever worn, or even manufactured.

Giovanni Borelli, who conceived many unworkable breathing devices, also designed an underwater suit which allowed a man "to walk on the bottom like a crab, or swim like a frog with his palms and web feet."

Benjamin Franklin, printer's devil who became a statesman, wrote that he used fins to

Rubber swim fins, sturdy and durable, are worn by millions of divers (right).

During the sixteenth century, Leonardo da Vinci focused a part of his genius on designing swimming aids. Below are drawings of his swimming gloves (A) and pneumatic life belt (B). Later, Borelli designed an underwater suit (C) with webbed feet so that "man could swim like a frog."

propel himself when he swam. "As a boy I made two oval pallettes each about ten inches long and six broad, with a hole for the thumb in order to retain it fast to the palm of my hand." Franklin said they aided his swimming but quickly tired his wrists. "I also fitted to the soles of my feet a kind of sandals, but I was not satisfied with them because I observed that the stroke is partly given by the inside of the feet and the ankles and not entirely with the soles of the feet."

The real advance in foot fins, in fact the first practical ones, were designed by Louis de Corlieu in the 1930s to help rescue aviators downed at sea. His secret was in using thin crepe-rubber plates glued over a thin blade

of flat spring steel. De Corlieu first marketed fins in 1935, and they spread quietly around the world among the diving subculture that crosses national and geographic boundaries. After World War II dozens of manufacturers sold millions of fins. All of them were made of molded vulcanized rubber, but none of them ever came close to matching the efficiency of the handmade sandwich fins built by de Corlieu.

Engineering has added another way for man to improve his propulsion underwater: either the so-called wet submarines or the electric scooters. Both are used by divers carrying breathing devices and are pulled or pushed by electrically driven propellers. The latest versions include the air cylinders, and the passengers, wearing only wet suits, simply grab the mouthpiece when they step in the sub or hang on to the scooter. In the wet subs the seating posture has been abandoned for the more logical "lying down" position. The advantages of these devices are increased speed and range, but they also reduce fatigue and lessen risk of the bends.

Electric scooters (above) and **wet submarines** (below), in which occupants lie down instead of sit, are some of the latest underwater propulsion units. Electrically driven propellers move the vehicles through the water.

Chapter VII. Longer and Deeper

If man was going to make much use of his new-found ability to dive, he had to cope with the reality of nature, once again, in exploring methods of penetrating deeper into the sea. Today science and technology are flirting with the very limits of deep diving with gas mixtures. Before we can even think about exploring new avenues of research, we still have to digest the lessons of the past and of recent years. It took centuries for man to move from the atmosphere into the sea, however slightly; hundreds of years of trial and error, theory and fantasy before the aqualung family of devices could be developed.

"The limits of oxygen and the danger of nitrogen demanded that a substitute breathing mixture be found."

But just as the density and pressure of seawater had once been a barrier, now the physical laws governing gases were throwing up a blockade to diving men. Henry's law on the partial pressure of gases took on special meaning in terms of life and death for divers. Decompression sickness was known from experience, then came nitrogen narcosis and oxygen poisoning. The toxic nature of oxygen came as somewhat of a surprise. Why should this gas, the breath of human life, prove so harmful when breathed under pressure? The complete answer is still being sought.

The shallow diving limits of pure oxygen and the dangers of nitrogen in compressed air demanded that a substitute mixture be found before diving deeply. The body can indefinitely tolerate oxygen up to a pressure less than 40 percent of one atmosphere. The answer, then, appeared to have a simple, two-stage solution: breathing gases must contain enough oxygen to support life, but not so much that it exceeds the allowable levels of partial pressure; and nitrogen had to be replaced with another inert or neutral gas which would not exert a narcotic effect on the body. However, the theory was much easier to arrive at than the practice.

The helium-oxygen diving era owed much to Captain A. R. Behnke, medical doctor in the U.S. Navy, and was inaugurated by Max Nohl's 520-foot dive in 1925. The inventive and brilliant Arne Zetterstrom demonstrated that hydrogen and oxygen could be used in deep diving. Later research was concentrated on helium. Navies were first to be concerned, because helium diving opened the door to military deep-salvage operations and to the rescue of the crews of sunken submarines. The navy organized research in experimental units of its own and gave contracts to universities. Offshore oil drilling created a demand for deep-diving specialists and industry joined in to finance experiments. In some cases the doctors later became divers, as with pathologist Guido Majno and physiologist Jacques Chouteau. And there were divers with a strong college and navy background, for example, Jean Alinat, probably more knowledgeable than any man in the world about diving and its effects on the human body.

The efforts of such men increased the knowledge that allowed for greater safety, as these volunteers risked their lives to make the wonders of the sea available to more men. It is immoral that research on decompression should be kept as an industrial secret.

Divers enter an **underwater decompression chamber.** *Once sealed, it is hoisted onboard ship where the divers' decompression process continues.*

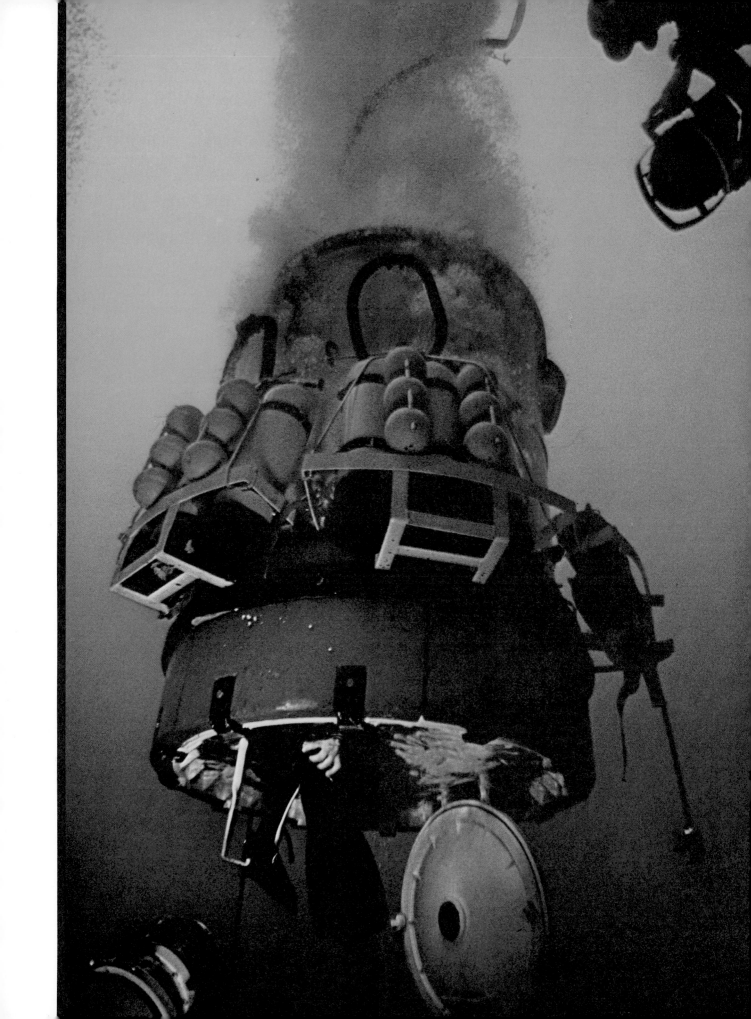

Physiologists monitor the vital functions of a test animal breathing pressurized gases. They hope to relate their findings to man.

Gas Hazards

Although man and his technology have greatly improved his diving equipment over the centuries, there are severe restrictions on the depth and length of time divers can spend in the world beneath the sea. This is not necessarily the fault of technology, but a fact of life concerning the human body and the property of the gases he breathes. Gases that are harmless at atmospheric pressure become toxic under higher and changing pressures. A quick review of the "gas hazards" is rather eloquent. The heavy, nearly inert nitrogen that makes up nearly 80 percent of air goes into solution in the blood and tissues. As we have seen, this nitrogen, when it comes out of solution as pressure decreases during ascents, forms bubbles in the bloodstream and blocks its flow, causing the bends. On deeper dives nitrogen induces an euphoric stupor called nitrogen narcosis, or rapture of the deep, which can incapacitate a diver.

Oxygen is the only gas really necessary for human life, but breathing pure oxygen can cause problems. For divers, the danger is very real, and it can become extremely hazardous to breathe pure oxygen at depths much greater than 20 feet. Breathed at high pressure, oxygen can cause sudden convulsions involving the central nervous system, while oxygen at moderate pressure breathed for long periods of time can irritate and damage lung tissue, an effect that had been foreseen by Joseph Priestly and Antoine Lavoisier as long ago as the eighteenth century.

Small laboratory animals breathing pure oxygen at 14.7 psi showed adverse affects within 24 hours and died within four days, showing capillary lesions and lesions of the alveolar membranes in the lungs. This adverse reaction, known as the Lorrain-Smith effect, is not as sudden in man, since his tolerance is greater than that of small animals.

Substantial partial pressures of oxygen can be used for dives of short duration because the lung irritation does not occur for several hours and usually disappears after a diver returns to normal breathing in the atmosphere. However, when these short dives are repeated daily, pulmonary disorders may manifest themselves after a lapse of several weeks. As a rule, it is better to avoid breathing oxygen under great pressure.

Despite all these hazards, sports diving with aqualung is practiced by millions around the world. Industrial deep diving with helium is widespread, and military divers in World War II and today were and are breathing oxygen under pressure and recycled through closed-circuit apparatus.

Sheep (right) as well as pigs make the best test subjects because their physiology is closely akin to man's. Breathing oxygen at high pressure has caused damage to the central nervous system and lung tissue of man and animals.

Partial to Danger

Tests performed on animals in 1878 by Paul Bert showed that oxygen breathed at high pressure acts directly on the brain and provokes an attack similar to that of epilepsy. The pressure at which this attack occurs varies with the time of exposure to pure oxygen, amount of fatigue, pressure of carbon dioxide, and individual susceptibility.

Translated into diving terms, the limit for diving while breathing pure oxygen is 15 to 25 feet. An epileptic seizure is not fatal in itself, but when a seizure occurs in water,

the victim drowns. However, for medical treatments in the dry, at rest, and under surveillance of specialists, there is not too much risk to submit a patient to three or even four atmospheres of oxygen for one or two hours maximum. If oxygen is diluted with a neutral gas in at least twice its volume, the effect is reduced; the acceptable partial pressure of oxygen increases, and for an exposure of a few hours a partial pressure of 2.3 atmo-

*Scientists study the **respiratory physiology** of this subject to learn how the body reacts to different levels and combinations of gases.*

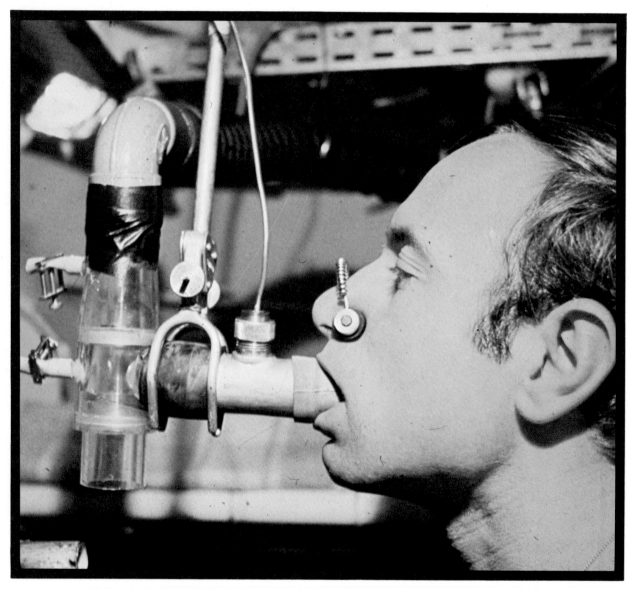

spheres is tolerable. This value is accepted for short-duration helium dives. With air, this critical partial pressure of oxygen is reached at a depth of 330 feet, corresponding to 11 atmospheres ($0.21 \times 11 = 2.31$). But, for extended stays, such as the days and weeks of saturation dives, it is not advisable to use mixtures and pressures where the partial pressure of oxygen exceeds double that of normal air, say 0.4 atmosphere. Practically, then, it is not wise to subject a diver to *prolonged* breathing with mixtures where the partial pressure of oxygen is more than half an atmosphere. Higher partial pressure can be used for short dives because pulmonary irritation does not have time to occur, and ordinarily such light irritation disappears rapidly upon return to normal conditions. Another gas that should never be found in breathing mixtures and must be avoided at all cost is carbon monoxide. This gas does not occur naturally in the atmosphere, but can accumulate in closed areas. It is a common product of internal combustion and is found where automobiles operate. Being present around motors, carbon monoxide can accidently enter tanks of compressed air when the intake valves of an air compressor are too close to an exhaust. Carbon monoxide is difficult to detect since it is normally odorless, but it is extremely harmful to humans since it reacts about 300 times as readily as oxygen with hemoglobin in the bloodstream which becomes unavailable for vital functions. Carbon monoxide present in a breathing mixture at about 100 parts per million will reach half its equilibrium saturation in about an hour. The acceptable level of carbon monoxide in compressed-air breathing mixtures is only 20 parts per million. Symptoms of carbon monoxide poisoning include rapid fatigue, red coloring of the skin, dizziness, and faintness. The only way to prevent carbon monoxide poisoning is to be sure of the source of compressed air.

This diver wears an **oxygen rebreather,** a device that recirculates exhaled oxygen and chemically absorbs exhaled carbon dioxide. The removal of waste products from breathing air is essential.

Finding a Substitute

The ideal breathing mixture appears to be a combination where nitrogen is absent and the partial pressure of oxygen kept as low as possible but still sufficient to sustain life.

Experiments have been conducted on various substitutes for nitrogen as an inert diluant for oxygen. Among inert gases, some are very heavy, such as xenon, krypton, and argon; they all showed narcotic effects much stronger than that of nitrogen. Neon has only a weak narcotic effect, and an acceptable solubility rate, but is very expensive. Two other gases, helium and hydrogen, are very light. Helium, on the one hand, is plentiful mainly in the United States and was fairly expensive since the government had a monopoly on it. Hydrogen is very plentiful and relatively inexpensive, but is dangerous to use since it can react explosively with oxygen when mixed in certain proportions.

*This experiment (above) has been designed to determine the effect that **breathing mixed gases under pressure** has on mental processes and visual perception. The results can be directly applied to deep divers breathing the same mixture.*

*Experiments using **pressure chambers** (below) to simulate water pressure are essential. If breathing mixtures being tested prove dangerous, they can be altered in the laboratory situation, but not when a diver is breathing them 1000 feet down.*

Any proposed mixture to be breathed by men must, of course, be tested on animals in compression chambers first. The first extensive testing of hydrogen as an underwater breathing mixture was done by Swedish engineer Arne Zetterstrom. Although a hydrogen-oxygen mixture is potentially explosive, the two gases do not react violently when the hydrogen is mixed with less than 4 percent oxygen. This percentage of oxygen, Zetterstrom reasoned, could not sustain life in the first 100 feet of the ocean, but when the pressure reached four atmospheres, the partial pressure of oxygen would be sufficient. His solution, then, was to breathe compressed air down to 100 feet, then change to a hydrogen-oxygen mixture. Caution must be exercised, Zetterstrom said, because "this cannot be done by a mere changeover since at the juncture where air and hydrogen combine, the mixture becomes an explosive. But if the air is first replaced by a mixture of 4 percent oxygen and 96 percent nitrogen, the risk of explosion is completely eliminated."

On August 7, 1945, Zetterstrom descended into the Baltic Sea with his three-stage plan to dive 100 feet on compressed air and then have his tenders at the surface cut off the air and switch to oxygen-nitrogen so that he could ventilate his suit and flush his system of excess oxygen. Then would come the changeover to hydrogen and oxygen and he could resume descent. He reached 528 feet breathing his unique mixture and suffered no apparent ill effects. Because of the new territory he was exploring, Zetterstrom had to devise his own decompression tables, and during a second dive, when he ascended to the 165-foot stop, the men hauling him up at the surface failed to allow him sufficient time to decompress. They continued pulling him up beyond the 100-foot mark, not even giving him time to change back from hydrogen-oxygen, which was too thin to sustain life near the surface. Zetterstrom passed out and died shortly after reaching the surface.

Though Zetterstrom's experiments proved the feasibility of using hydrogen-oxygen mixtures, the dangers and his death—even if from unrelated causes—steered all future diving experiments away from hydrogen. Recent laboratory experiments (Brauer and others) on animals tend to indicate that with hydrogen-oxygen mixtures a narcosis develops from 825 feet downward, which makes helium the more advantageous.

*Decompression is still a problem that all deep divers must face. At one time stage decompression was the only ascent method available and it was time-consuming. Now divers can enter **pressurized chambers** and be hauled onboard ship where decompression inside the chamber is continued in safety.*

The Helium Trials

As soon as the hazards and limits of deep diving with compressed air became known, research turned in other directions.

As early as 1917, Elihu Thomson proposed helium as a substitute for nitrogen. Then in 1924 a group of chemists, physicists, and medical doctors, including Behnke, Yarborough, Yant, Sayers, and others working for the U.S. Navy, began experimenting with helium as a breathing gas. Helium, which is

Hannes Keller left this chamber at 1000 feet and swam free, breathing gas from a hose.

seven times lighter than air, was originally used to fill balloons and dirigibles. When used in breathing mixtures, it was found that there was no effect similar to nitrogen narcosis and, as a result, a few years later Nohl made the first actual deep dive using a mixture of helium and oxygen. In 1935 a series of safe dives were being made to depths of about 400 feet. The success of these early experiments proved the feasibility of helium mixtures, and soon almost all industrial dives deeper than 200 feet were made using the helium-oxygen combination.

Hydrogen, the lightest element in the universe, was a possible component for a breathing recipe. But Arne Zetterstrom's death, and the caution with which hydrogen must be handled, apparently deterred other divers.

Sixteen years after Zetterstrom's experiments, a young Swiss mathematics teacher named Hannes Keller dived 728 feet down into Lake Maggiore in Switzerland, breathing pure oxygen near the surface, then switching to mixtures of his own devising, using three different cylinders of breathing gases in all. After testing his method to 1000 feet in the French navy's hydropneumatic chamber in Toulon a year later, in 1962 Keller was lowered a thousand feet in a special turret, entered the water through a bottom hatch, and swam free with an air hose in the deepest penetration of the sea an unarmored diver had ever made.

Laboratory experiments on men have shown that there are no psychological problems that develop when using a helium-oxygen mixture down to at least 1300 feet. With animals, experiments have been successfully completed to more than 3300 feet, equivalent to 100 times atmospheric pressure. The low density of helium is of considerable advantage in facilitating gas exchanges and pulmonary ventilation. Studies such as those at the Center for Advanced Marine Studies in Marseilles, France, have demonstrated that the limits of human diving with heliox are in the vicinity of 2000 feet. They also have shown that it is the density of breathing gases which causes problems in experimental animals at simulated depths of 2600 to 3300 feet. These animals experience the same difficulties breathing helium-oxygen at these depths as animals breathing nitrogen-oxygen at 550 to 700 feet.

But even helium has shortcomings. The hope for shortened decompression times, due to faster diffusion out of the system, for example, is negligible in tests conducted at 225 to 300 feet. Helium mixtures require sophisticated, expensive equipment and complicated manipulations. Helium is a good conductor of heat and exacerbates body heat loss. Even insulating materials are permeated by gas and lose most of their effectiveness.

Keller prepares for his 1000-foot dive in 1962. His was then the deepest penetration of the sea that had ever been made by an unarmored diver. He used three different cylinders of breathing gases.

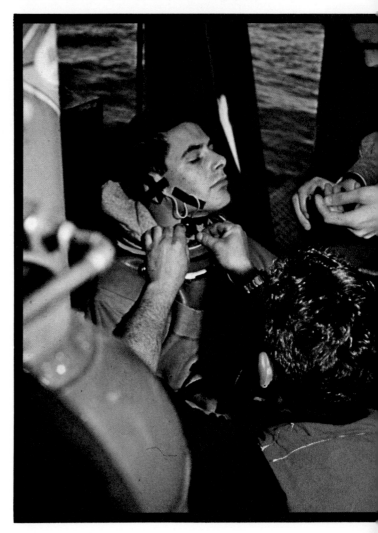

Advances

The conventional open-circuit, self-contained compressed-air breathing apparatus has opened the world of the sea to millions of people. Of all the breathing devices available it is by far the easiest and safest for the novice diver to use. But in addition to the restrictions imposed by oxygen and nitrogen in a compressed-air system, there are limitations in the device itself. It is bulky, and out of water it is quite heavy—about 50 pounds. The valve that releases air from the compressed-air cylinder operates noisily underwater, where sound is conducted much better than in air, and the exhalations of the diver are not only noisy, but the bubbles are conspicuous enough to frighten some marine creatures. And because of the open circuit, a large volume of gas is used in relatively short periods when the diver is at great depths. A big two-tank unit is good for nearly two hours of work at 30 feet, but less than 30 minutes of light work at 200 feet, excluding decompression time, which for such a dive would total an hour and a half.

To extend the capability beyond that of the open-circuit breathing apparatus, it would be necessary to recycle the neutral gas, since it is not contaminated and does not need to be vented. Oxygen must be provided at a partial pressure of between 20 percent and 40 percent of one atmosphere. Exhaled carbon dioxide must be removed from the system in order to keep the diver healthy.

This challenge has been met almost simultaneously by several manufacturers. Among their devices are Beckman's Electrolung, Westinghouse's Krasberg unit, and General Electric's Mark 10. They generally use helium as neutral gas. The breathing mixture is automatically controlled and monitored electronically so that the diver knows if the oxygen partial pressure varies from tolerable limits. Oxygen is automatically introduced by an electrovalve operated by the sensors. Carbon dioxide from respiration is scrubbed in a chemical absorbent canister. Electric lungs can be operated manually in emergencies and in later stages of decompression where a changeover to pure oxygen can be accomplished. The units operate silently and leave no bubbles. They have been tested to 1650 feet in pressure chambers and to 350 feet in the sea. When the gas tanks are filled to capacity, the diver can remain underwater from four to six hours at any depth, provided his decompression is made with the help of a chamber or of another unit.

The most important pieces of equipment in such closed-circuit systems are the sensors that measure the partial pressure of the oxygen in the breathing bags. They are basically polarographic sensors, generating an elec-

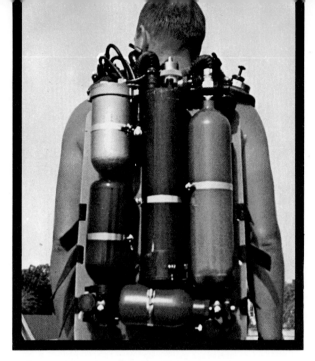

tric current transmitted to a meter worn on the wrist like a watch and to the electrovalves which automatically regulate the oxygen flow keeping it at tolerable levels.

These advanced closed-circuit systems have allowed longer and deeper stays in the sea, but these forays, however successful, have been only for relatively short periods of time. The systems still carry some risk. The increased capabilities allow a diver to get into trouble by going too deep or staying down too long. Therefore these devices must remain, for the time at least, professional or military tools. Aqualungs are still the only systems that leave no doubt about the exact nature of the gas that is taken in.

*A diver emerges from frigid polar waters (opposite) wearing General Electric's **mixed-gas rebreather**.*

The Krasberg unit *(above) is one of several systems that electronically control breathing mixtures.*

*Diver (below right) is wearing an **Electrolung**. A similar unit is the **Mark 10** (below left).*

Chapter VIII. Living Underwater

In his effort to cheat nature and function underwater, man has encountered great difficulties. He has surmounted many of these by using artificial breathing mixtures and special diving systems. New difficulties, such as helium's distortion of the voice, are still to be resolved.

Man has had to cope with an inflexible physical law of nature: breathing gas under pressure requires long and often laborious decompression. There is considerable interest in methods to improve the diver's efficiency, the ratio between working time and wasteful, painful, dangerous decompression time. As a diver's stay at a given depth increases, the time required for decompression rises rapidly until a ceiling is reached. This is because once the body has become saturated with neutral gas the quantity to be eliminated during decompression remains the same. With helium, which is used in deep dives to avoid the hazards of nitrogen narcosis, the saturation point is reached in a few

---

*"Diving efficiency is
greatly increased by staying
down for several days."*

---

hours. In compressed-air breathing, the saturation stage is reached only after longer stays because the neutral gas, nitrogen, is less diffusive than helium.

Because of the problems involved with a decompression starting from a state of saturation, it is obvious that reasonable productivity in great depths can only be obtained by staying down for several days. This finding intrigued Captain George Bond and led him to study saturation dives for five years. Test animals showed that saturation dives

could be made without physiological risks. On September 6, 1962, Robert Stenuit stayed 24 hours in a small, submerged recompression chamber at a depth of 200 feet in the Bay of Villefranche. He breathed a helium mixture and dived several times out of his cocoon using a breathing tube. From September 14 to 21, Albert Falco and Claude Wesly lived in Diogenes, our Conshelf I station, which is 35 feet deep; they worked several hours a day at depths to 85 feet.

These first two successful experiments, together with the 1000-foot dives of Hannes Keller, led to the establishment of various habitats, or undersea living stations. In September 1963, Bond and a U.S. Navy crew organized a successful simulated saturation dive, "Genesis," at 200 feet. In 1964 the four-man Sealab I was placed at the same depth in the sea. Our Conshelf II had already completed its month-long stay beneath the Red Sea, and in 1965 underwater habitats successfully reached 328 feet in Conshelf III and 430 feet in Man-in-the-Sea II.

These successes were followed by the industrial development of the cachalot system in which restoration divers are transferred to and from the surface vessel in pressurized capsules. Then lock-in-lock-out (LILOS) submarines increased the efficiency of an exploration sub. The final step will be a larger submarine conceived to serve for more than a week as bottom-tender for true saturation divers operating down to the very limit of helium diving—2000 feet. This is what the *Argyronete* is about (see pages 130-131).

*The psychological well-being of **Conshelf II aquanauts** is maintained by providing the comforts of home. Here, two men relax over a game of chess.*

Underwater Colony

It was an unlikely spot to set up a camp, if there is a logical location for an underwater habitat, but we wanted an ocean environment which would present problems. If man could cope when the going was rough, then sea bottom settlements would be all the more possible in more hospitable waters. So we settled on Sha'ab Rumi, the Roman Reef off the Sudanese Coast in the Red Sea. The main unit was 32 feet below the surface on the terrace of an underwater cliff, with the satellite Deep Cabin 82 feet down.

There had been previous attempts to stay submerged for prolonged periods. In 1962

Robert Stenuit stayed one day at 200 feet, in and out of an aluminum chamber built by Edwin Link. Shortly afterward, Albert Falco and Claude Wesly remained in Continental Shelf Station Number One for a week. This station was on the bottom 35 feet down, and they worked for up to five hours a day at depths to 85 feet. But for Conshelf II, we were planning a true colony, where five men would live for a month, in addition to the two men who would inhabit the Deep Cabin for a week. The Conshelf II complex included a main unit we called Starfish House because of the branching workshops and living quarters off the main chamber; a wet garage for the scooters and tools; and a dry domed hangar for the diving saucer. The various buildings were linked together with telephone and television lines, and cables from the surface ship *Rosaldo* provided power for communications and electrical equipment and the compressed air which kept the pressure inside at two atmospheres. Since much of the research work was going to be carried on outside Starfish House, sometimes as far as 260 feet down, we ringed the area with antishark cages in which the divers could seek refuge and find emergency air tanks.

An important part of the operation—and in many ways the most vulnerable because of the possibility of surface storms—was *Rosaldo,* where we had duplicate sets of electrical generators and air compressors. Among the crew on the surface were Falco, pilot of the diving saucer, and his two mechanics; doctors; and a dozen service divers who ferried supplies to the oceanauts below.

The project was only six months in the planning, and though the men were pioneers, we didn't overlook some things that could make them more comfortable. They ate fresh fruit and cheese, drank wine and good cognac, had chess sets, books, and a high-fidelity tape player system to entertain them in their leisure time, as well as ultraviolet lamps to keep their skin healthy.

The five-man crew in the habitat was headed by marine biologist Raymond Vaissière. The others were cook Pierre Guilbert, the oldest at 43, mechanic André Folco, diarist Pierre Vannoni, and Claude Wesly, a diving veteran though he was only 31 years old.

Divers (opposite) hang from the entrance ladder of Diogenes (Conshelf I). *Oceanauts* (above) set up black-and-white cubes as part of an experiment to test their mental alertness.

Captain Cousteau visits oceanauts Falco and Wesly (below) within the confines of Conshelf I. The unit was 35 feet below the surface.

At Home in the Sea

Conshelf II was primarily an experiment. We wanted to test man's capacity for prolonged living on the sea bottom and mainly his efficiency when working extended hours from a pressurized habitat. The oceanauts' psychological stamina was as important as their health. They were examined each day by our diving doctor, Jacques Bourde. The Starfish House men quickly tired of his daily visits, protesting their good health, but the medic told them simply and firmly, "I must record data for the use of other sea floor establishments." Soon the men living in the Conshelf station realized there were subtle effects of the pressure. Beard growth was retarded. Small cuts and abrasions healed much faster than similar injuries suffered by our other divers living on the surface.

The first serious medical problem was earaches. All the men complained of burning, stinging sensations in their ears. But soon the surface divers were voicing the same com-

Starfish House, already in the water, and the *diving saucer's hangar* are made ready for their epic adventure below the waves (above).

Captain Cousteau and aquanauts of Conshelf II (opposite, top) relax over some champagne, but high cabin pressure keeps the bubbles in solution.

Conshelf divers (below) swim off in the direction of Starfish House. Diving saucer (below right) enters its submerged hangar.

plaint, and we then knew it had nothing to do with the habitat. The cause, it turned out, was a mass of almost invisible siphonophores which inflicted venomous stings on the most vulnerable part of the divers—their ears. The solution was simple—ear washing was dropped from the daily routine and ear wax was allowed to form a protective coating.

Potentially more important than the medical problems were the mental ones. How would five relative strangers adapt to living in close, if comfortable, quarters where the constant humidity and temperature were high enough to induce discomfort and nasty tempers. Once an air conditioner was installed, this possible trouble source was shut off. During the first few days, of course, all the men were excited at the thrill and challenge of the novel tasks that lay before them, and certainly zoological observation in a beautiful coral reef was not too tedious a job. However, when routine set in, we expected trouble (if, in fact, it was to come). But the period of adaptation passed without serious incident, and Pierre Vannoni wrote in his diary, "I am beginning to be aware of time passing, which I had rather forgotten about. I feel I may rise to the surface next week without having seen

and experienced absolutely everything." Vannoni was an interesting case: he had been buried alive by an explosion during the war and had suffered claustrophobia ever since. But in Conshelf his fears vanished; he also found that once his routine was established, he slept more deeply than usual and found it difficult to wake up.

One of the unpleasant experiences came midway through the stay when I took down some bottles of champagne to toast the milestone. In the high pressure of the habitat, the bubbles failed to come out of solution; the wine was as still as the world outside.

The Deep Men

An important part of the Conshelf II plan was the Deep Cabin, where Raymond Kientzy and André Portelatine were to spend a week in a helium-oxygen atmosphere. They were to leave their rocket-shaped home 82 feet below sea level on work trips while breathing compressed air from aqualungs. We reasoned that they could dive regularly to 165 feet from their base and take short trips down as far as 330 feet without too great a risk of nitrogen narcosis.

Deep Cabin was equipped with two chambers: the lower one, called the wet room, containing the diving gear, tools, and sea hatch; and the relatively dry living quarters above. It was not as spacious as the tile-covered, pastel-colored Starfish House, but there were bunkbeds, a kitchenette, and the full range of communications equipment.

One piece of apparatus that was missing was an air conditioner. We had originally planned the experiment for March, when we thought the water at that depth would be sufficiently cool to make Deep Cabin livable. But in July, the Red Sea outside Deep Cabin was 86° F. and the humidity was always 100 percent.

There were also equipment malfunctions— in the telephone, the refrigerator, and the backup oxygen system. Living in a 24-hour sweat bath also caused Kientzy's and Portelatine's appetites to malfunction. But Dr. Bourde visited them every day and said that other than being uncomfortable the two were in good health.

The men continued to "perspire like fountains," as Kientzy put it, and often took swims just to wash off. Portelatine suffered from the inflamed ears that the others experienced, but his pain didn't deter him from

Deep Cabin (opposite and above left) was a part of the Conshelf II project. It was set 82 feet below the surface of the Red Sea.

A Deep Cabin diver (above, right) studies the reactions of fish to colored light. *Portelatine and Kientzy* (below) rest in the living quarters where humidity was 100 percent.

his chores. Their main objective was to dive to 330 feet from the helium atmosphere base while breathing air. At that depth, the pressure is 11 times that of the atmosphere at sea level—161.7 psi.

On the sixth day of their stay, Kientzy and Portelatine descended toward the edge of the reef on which Deep Cabin rested. Then they dropped to 300 feet, past hundreds of garden eels anchored in the substrate, swaying in the current, feeding on the plankton. Down the deep men went, toward their goal of 330 feet. They felt no drunkenness, no rapture of the deep that ordinarily accompanies air-breathing men at that depth. And though they entered their achievement in the log as 330 feet, they later revealed that they had reached 363 feet—a record for the deepest compressed-air dive in history, but one which they did not want to claim because setting records was not their purpose.

113

In Bermuda Waters

The U.S. Navy's first underwater habitat was Sealab I, the brainchild of Captain George Bond, a diving physiologist, who was the first to propose the principle of saturation diving. Every diver knew that the longer a person remained in the water at a given depth, the longer the decompression time would be on the return trip to the surface. But Bond reasoned and subsequently demonstrated that once the body absorbed all the inert gas it could—in other words, became saturated with either nitrogen or helium according to the mixture breathed—the decompression time no longer increased. It made no difference if the diver were down a day, a week, or a month.

The Sealab I habitat was tubular, with water ballast in the end sections and living quarters in the center of the 40-foot-long cigar. It was 10 feet in diameter, thus the four aquanauts could stand up. Sealab I was lowered by a crane to the Plantagenet Bank 26 miles off Bermuda in the Atlantic and settled at a depth of 193-feet below sea level. Support lines, attached to a navy floating platform, the YFNB-12, kept the atmosphere inside at 80 percent helium, 16 percent nitrogen, and 4 percent oxygen. Temperature was 90° F. and relative humidity 70 percent. The four aquanauts were all navy personnel: Robert Thompson, Robert Barth, Lester Anderson, and Sanders Manning. When they left their base, they wore semi-closed-circuit oxygen rebreathers. Manning's apparatus malfunctioned while he was on a photography mission and he managed to make it back to the sea hatch before losing consciousness. Anderson was able to snag Manning's body just as it was beginning to drift away. Anderson got him breathing again within a couple of minutes and despite ruptured blood vessels in his eyes, Manning completed the 11-day stay on the sea floor.

In reviewing the accomplishments of the Sealab I project, Bond said it took about five days for the men to settle down to routine—just as in our Conshelf II experiment. But once were ready to work, there wasn't enough to do. "In this connection," Bond said, "I am sorry to say that the marine biologists who were promising such input into the program have not been forthcoming.... So here we sit in a biological paradise 24 hours a day, and damn few tools to work with." The stay was shortened because of the threat of a hurricane. The achievements of Sealab I were significant.

Sealab I (left) *is loaded onboard ship. It was the U.S. Navy's first underwater habitat.*

At a depth of 193 feet, **Sealab I** *(right) was home to four aquanauts for 11 days of scientific study.*

Navy Tries Again

Sealab II provided an interesting contrast in man's ability to penetrate outer and inner space. Astronaut Scott Carpenter knew what it was like to hurtle through space miles above the earth, but as part of the Sealab II crew, he would spend 30 days on the sea floor 205 feet below sea level, and about as much training was desirable in either case.

The Sealab II habitat was larger than its predecessor, 57 feet long and 12 feet in diameter and less tapered at the ends—resembling a railroad tanker car. The plan was for 28 men to stay down in 15-day shifts, Carpenter and navy doctor Robert Sonnenberg remaining down for 30 days, so that at any given time during the 45-day stay from August 28 to October 14, 1965, there were 10 men living in Sealab II. The project was conducted in the Pacific, where there was no danger of hurricanes, less than a mile off the coast of southern California.

Collecting all physiological data from the inhabitants was one of the primary goals of the project, and each day the aquanauts sent up samples of breath, blood, urine, and saliva in a pressurized capsule. In addition to chores like placing monitoring instruments on the sea floor, the men performed strength and manual dexterity tests before and after each dive. Every evening there were daily activity and mood checklists to be filled out and occasional "brain teaser" quizzes and arithmetic tests to determine the possible effect of the high-pressure helium-oxygen atmosphere on the brain. The breathing mixture was 77–79 percent helium, 18 percent nitrogen, and 3–5 percent oxygen. Lithium hydroxide was used to absorb carbon dioxide. A test on the effect of drugs under pressure came when Carpenter and another aquanaut, Billy Coffman, were stung by venomous scorpionfish. The men responded to the antihistamines and pain killers, and the swelling and pain subsided in 24 hours.

Sealab II (below) *is larger than its predecessor, measuring 57 feet long and 12 feet in diameter.*

The U.S. Navy's **Sealab II** *(opposite, top) is towed into position off the southern California coast.*

A Sealab II engineer repairs a damaged headset (below) as queenfish look on.

Two Sealab II aquanauts (left) emerge from their craft after spending at least 15 days below the surface of the Pacific.

An experiment was conducted with Tuffy, a porpoise trained to respond to sound signals. At first he was shy and irresponsive, but then rose to the task by delivering mail between the surface and Sealab II and by finding men outside the habitat who signaled that they were in danger and acted accordingly. Working conditions in the water were made difficult by the poor visibility and cold, although heated suits allowed men to double their stays in the 50° F. water to two hours.

The bodily changes that the aquanauts experienced were transitory: some decrease in strength, manual dexterity, and coordination were observed. These conditions abated when the divers returned to the surface after their 56-hour decompression. There was no measurable change in mental test results before and after the stay. Humidity, as always, was a problem. But as spaceman Scott Carpenter later said, it was "the most richly satisfying experience of my life."

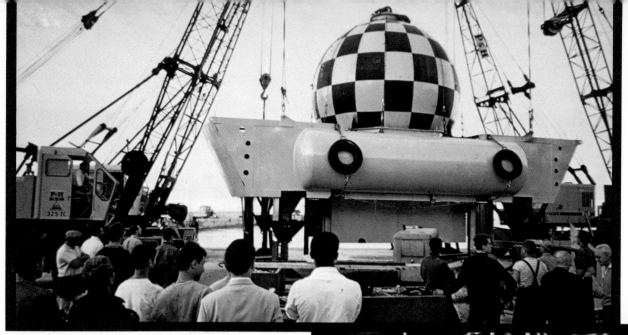

Cutting the Cords

While aquanauts of Sealab II were down on the floor of the Pacific, we were 6000 miles away in the Mediterranean working on Conshelf III. This was a radical departure for us, since the habitat would settle 328 feet below —too far for support divers to reach it from above. If man was going to inhabit the continental shelf, he had to sever the umbilical cord to the surface, and this was a decisive step in that direction.

Conshelf III consisted of a spherical chamber 18 feet in diameter resting on a chassis 48 by 28 feet. An engineering characteristic of the habitat that proved essential for safety was that the sphere could resist inside as well as outside pressures higher than the 11 atmospheres to be met at the operating depth. The men were breathing a mixture of 98 percent helium and only 2 percent oxygen. Under 11 atmospheres, this was just a little more than at the surface. There were no ill effects on the divers' bodies, but some of the machinery broke down in the lighter medium, even after the many tests we had made before. Helium is so light that the oceanauts' vocal cords vibrated much faster than they do in air. The result is a squeaky, high-pitched voice that sounds like Donald

Duck's. The helium voice—a condition that affected Sealab and any helium diver—made communication difficult between the men below as well as on phone calls to the surface. Voice from topside or music from loudspeakers were, of course, undistorted.

Conshelf III (opposite), set 328 feet deep, was an almost self-sufficient habitat. It relied on the surface only for power.

Aquanauts (below) prepare to enter Conshelf III. *Divers inside Conshelf III* (right) check some of their equipment after returning to the habitat.

World Without Sun

While working outside Conshelf III, the oceanauts were tethered to the habitat with 200-foot breathing hoses that pumped heliox through one tube and sucked exhalations through another; the foul air was scrubbed in the habitat's recycling system where the helium could be used again. The oceanauts also wore aqualungs filled with heliox, good only for a few minutes, in case something went wrong with the breathing hoses.

The major task of Conshelf III divers was to prove that men could perform practical work, such as the difficult maintenance of oil rigs—Christmas trees as the rig monkeys call them—at a depth twice as far down as had previously been accomplished. Though there was no oil well below our Christmas tree, which actually rested 370 feet below sea level, the oceanauts accomplished such chores as changing valves and setting up equipment. That made believers out of skeptical oil engineers watching our TV monitors at the surface. One of the oceanauts, Christian Bonnici, worked seven straight hours, an unheard-of achievement at that depth.

Among the things Bonnici accomplished was threading a stiff wire through a series of pressure-proof seals—a task considered almost impossible underwater.

The entire system was threatened when a violent storm almost destroyed our "shore to station" control installations and lines. The oceanauts remained completely unaware of the drama.

The three-week stay of Conshelf III reaffirmed our experience that surface vessels and research equipment are more vulnerable to damage than the underwater habitat and that machines fail more often than men.

Conshelf III (*opposite, top right*) *rested over 300 feet below the surface and proved man could perform difficult maintenance jobs at that depth (opposite, top left; opposite, bottom).*

Captain Cousteau (*above*) *directs the activities of Conshelf III from a lighthouse on Cap Ferrat.*

The crew of Conshelf III *enjoys dinner and some time off from their work (below). The habitat remained on the bottom for three weeks.*

In the Cause of Science

The Tektite project was a cooperative venture of government, industry, and educational institutions, which gave scientists the opportunity to study the ocean on a participant-observer basis.

The divers and engineers of Conshelf, Sealab, and Man-in-the-Sea programs had solved all the problems for this seven-month routine operation where more than 40 scientists took turns living in teams 60 feet below the surface on a reef in Great Lameshur Bay off St. John Island in the U.S. Virgin Islands. The Tektite II habitat, named for the glassy meteorites which occasionally shower the earth and are found strewn about the sea floor like marbles, has a square blocklike base with two vertical towers 12½ feet in diameter and 18 feet high. Support lines, the umbilici, ran to a land center 200 yards ashore.

Accommodations included hot and cold running water, internal temperature of 80° F., and relative humidity maintained at 40 percent. When the scientists ventured outside the habitat, they eventually used the

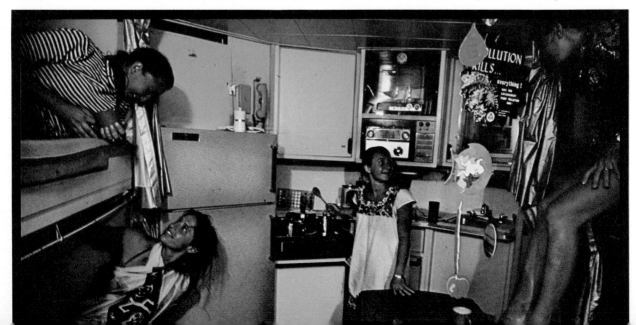

advanced, electronically controlled closed-circuit breathing apparatus developed by General Electric, one of the principal industrial sponsors of the project. The silent lungs allowed divers to approach the ocean creatures more closely. The primary concern of all the scientists was the ecology of the reef community.

One of the more unusual aspects of the Tektite program was that one of the rotating teams of scientists were all females. Simone Cousteau, my wife, had lived in Conshelf II for a number of days, but never before had women "manned" an underwater habitat.

Sylvia Earle, a marine botanist, was the leader of the crew of five, which stayed below for 14 days. She felt the undersea station offered a total work-study environment. If there was a complaint about the quarters themselves, the scientists said the engineers didn't incorporate enough laboratory space into the Tektite II design. Earle concluded, "To me, the next logical step is to move our laboratories to the environment itself."

Tektite was followed by a proliferation, all around the world (United States, Japan, Great Britain, Italy, USSR, Germany, Canada, Bahamas, Netherlands, Cuba), of simplified, inexpensive, and often crude undersea stations, which lacked the controls and complicated equipment of the experimental units. All of these stations are in shallow depths (25 to 50 feet).

Joseph MacInnis lowered Sublimnos 30 feet down in Lake Huron, Ontario. This first freshwater habitat accommodates two to four divers. Hydrolab is another venture located 50 feet down in the Bahamas. Capable of accommodating three or four divers, it has two remarkable features; it receives electricity, compressed air, and fresh water from an unattended buoy, for up to seven days; and it can be used as an atmospheric pressure so dry transfers can be made between it and Perry's *Shelf Diver* submarine.

Aquanauts of Tektite II (opposite, above) swim off to investigate a reef community. A crew of five women "manned" the habitat for a 14-day period (opposite, below), the first time an all-women team took part in an underwater habitat project.

Hydrolab (right) is another habitat in the Bahamas. It can be operated at atmospheric or ambient pressure. Perry's submersible pressure chamber (above) is a classic transfer capsule.

A cachalot diving system is used in the Gulf of California from this 300-by-80-foot work barge. The main decompression tank stays on deck.

Industry Adapts

Saturation diving had demonstrated that scientists, engineers, and other undersea workers could spend their time either on the job or at rest in an underwater habitat without having to decompress for hours after each day's efforts.

Industrial divers, though, found that in many cases habitats were impractical or unworkable for such reasons as expense or location. As a result, the cachalot system was developed for divers at dams and underwater construction and offshore oil well sites. The divers are put down with a special submerged decompression chamber, usually referred to as an SDC, which is lowered from a surface slip; the divers reenter the SDC at the completion of their work shift or at the end of a day's work. The SDC is then raised either to a dock or the deck of a surface ship and the divers are transferred through an airtight lock into a more comfortable pressurized chamber, where they live between work shifts under the same pressure as the bottom. There are now dozens of such transfer-under-pressure systems. Though all these systems are based on the same principle, there are wide variations in size and in technical details.

These systems allow saturation diving to a depth of 1000 feet and maybe soon to 2000

The cachalot system has obvious advantages, at least on paper: possibility to move from one site to another, to hoist up and leave when a storm is announced, and to keep the divers on deck under easier medical control. True. But it has serious drawbacks. The tender ship has to be large and specially equipped, its cost will be running all year, even if it is sparsely used. And handling a heavy weight onboard a ship that is pitching and rolling most of the time has always been a hazard. Also, helium divers need a lot of local support, gas, heat, and comfort, and a cramped SDC is far from affording such necessities. The list of casualties in the past few years is unfortunately a long one.

A submerged decompression chamber, or SDC (below), is raised from the depths. Its occupants will be transferred through a pressurized lock to a more spacious and comfortable chamber onboard ship.

Pressurized elevators (above) allow divers to use the time-saving saturation-diving technique without having to live in underwater habitats.

feet, by teams of two, three, or four men. The work can be done in relays, for while one team is resting topside, another crew can descend and continue working. While they are under, the divers generally use breathing hoses attached to the SDC, which in turn is replenished from a self-contained gas supply or from lines connected to the surface tender ship. Such "surface deep stations" are commonly built and used for offshore oil production by Japanese, French, American, British, and Australian firms and navies. The original cachalot used such advanced accessory equipments as the Krasberg electronically controlled breathing units and O'Neill's hot water units.

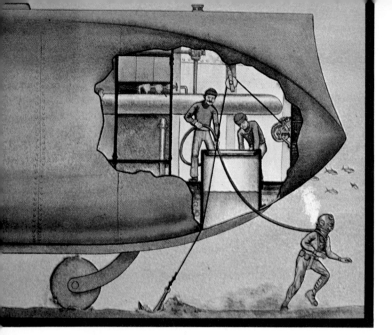

LILOS

Lock-in-lock-out submarines (LILOS) are undersea crafts with an egress hatch allowing divers to leave and perform tasks that clumsy manipulators would be incapable of doing., LILOS are a compromise between vehicles and habitats. They are too small to carry the energy sources, gas, and instrumentation necessary for deep saturation diving. Actually the idea is much older than saturation diving or habitats. Jules Verne wrote about the *Nautilus* in 1869, and 25 years later Simon Lake built the *Argonaut Junior*.

The wooden *Argonaut Junior*, propelled by three hand-cranked wooden wheels, was equipped with an air lock permitting helmet divers to walk out and work. Lake later built a 36-foot iron version, called simply *Argonaut*, launched in 1897. In the mid-nineteenth century William Newton built a vehicle with no hatch—the diver sat on a platform outside the craft, drawing air from inside the vessel, which in turn was fed from a surface ship. The *Bateau-cloche*, built in 1844 by a Dr. Payerne, included a diving bell; the same inventor later constructed a larger LILO called the *Pyrrhydrostat*. In 1899, L. de Riguad devised one of the largest such submersibles, a lemon-shaped 80-foot-high hull divided into six compartments. The craft could settle on the bottom and sit on its four legs. In an emergency the uppermost compartment was detached, and the crew ascended in that section alone.

Modern LILOS include Perry-Link's *Deep Diver*, I.U.C.'s *Beaver*, Smithsonian's *John-*

Simon Lake built the **Argonaut** *(above) in 1897. It had an air lock that allowed helmeted divers to move in and out of the craft.*

Perry's submersible **Shelf Diver** *(below left) is a lock-in-lock-out submarine (LILOS) that can carry four crewmen down to depths of 800 feet.*

Johnson-Sea-Link *(below right) is a modern underwater vehicle that has a transparent acrylic hull to facilitate observations below the surface.*

son-Sea-Link, Perry's *Shelf Diver*, and International Hydrodynamics *Pisces III*, which is operated by the Canadian Department of Defense. Others are in construction in various countries around the world.

Imagine a ruptured pipeline somewhere underwater. The oil flow has been shut off, and the damage has been located by sonar or by magnetometer. A lock-in-lock-out submersible is launched from a mother ship in the vicinity of the rupture. The pilot takes the LILOS down and sets his craft on the bottom near the damaged pipe. Once their compartment has been pressurized to match the depth of the water outside, the divers grab their tools and head to work. After the repair is finished, the divers return to the LILOS and begin to decompress. If there is a need for prolonged decompression or medical attention, the diving chamber of the LILOS is able to be mated with a recompression chamber on the deck of the mother ship.

The *Johnson-Sea-Link* was launched in 1971, but two years later the vessel became entrapped in the hulk of an aged destroyer off the coast of Florida. The craft was recovered, but the lives of half the crew had been lost. Veteran divers Albert Stover and E. Clayton Link, son of the inventor, died from the cold. Their recirculating breathing system's capacity had been exceeded, and they were forced to use open-circuit breathing, raising the pressure of their compartment until they had to use helium mixture without any heating appliance.

The Perry *PC 15,* like the *Sea-Link,* is equipped with a transparent acrylic hull to increase the field of vision.

Beaver IV is a LILOS used by industry and the scientific community alike. Divers can leave the craft when necessary and perform tasks outside that internally operated manipulators could not.

Have Habitat, Will Travel

Mobile homes are a way of life for many Americans, and gypsies in Europe have been using them for centuries, so it was only a matter of time until someone built mobile undersea habitats. The most mobile, or rather portable, is Edalhab, designed by students at the University of New Hampshire. It is capable of making short stops in chosen places; it is then picked up by the mother ship, *Lulu*, and transported to a new site.

Edalhab is small, 8 by 10 feet inside, and is tethered to *Lulu*, the 105-foot-long tender operated by the Woods Hole Oceanographic Institution. Edalhab was used in the three-month Florida Aquanaut Research Expedition (FLARE), whose aim was to study the geology, evironment, natural history, and ecology of coral reefs. The project, sponsored

Edalhab, with tender ship **Lulu** *(above) and on the sea floor (below), is a portable habitat.*

by the National Oceanic and Atmospheric Administration, included seven scientific teams who required—and at the same time tested—a movable habitat. Edalhab, anchored at a depth of 45 feet for each project, was filled with compressed air at ambient pressure. As a result, there was a four-day limit on each mission so that the problems associated with breathing high-pressure oxygen for a prolonged period could be avoided. In other extended stays in habitats, the normal 20 percent oxygen in air is reduced in the breathing mixtures, so that oxygen partial pressure remains under 0.4 atmosphere.

Scientific experiments included observing fish reactions to traps; comparing a healthy coral reef and one in polluted waters closer to Miami Beach; measuring the rate of coral reef growth; studying seaweed and other marine plants; implanting an artificial reef of old tires; and taking a fish census.

Whereas Edalhab is portable, another habitat, Aegir, is mobile. It is designed to descend and surface by itself, as did Conshelf III in 1965. Built and operated by Makai Range, Inc., of Hawaii, Aegir is capable of housing six persons for two weeks at a maximum depth of 580 feet. But although the breathing gas cylinders are mounted on Aegir itself, it is not self-contained: a surface ship, *Holokai,* provides power via an umbilicus. There is also a backup battery aboard the habitat in the event that power from the surface should fail. On the bottom, the breathing mixture is 91 percent helium, 7.2 percent nitrogen, and 1.8 percent oxygen.

LILOS as well as mobile undersea homes, such as those in existence today, increase man's perception of what the real problems of undersea living are. But they can only go halfway because the vast funding necessary for such enterprises is not yet available.

A diver stows gear in Edalhab. Aquanauts inside breathe compressed air at ambient pressure.

Edalhab was used as an underwater base in the FLARE project to study coral reefs around Florida.

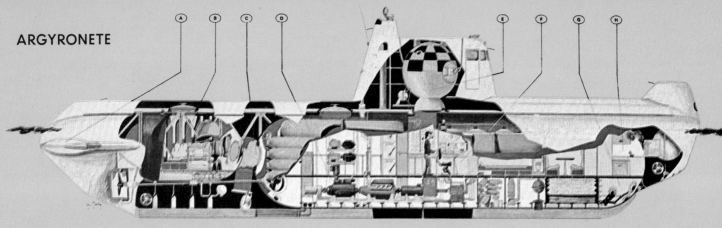

ARGYRONETE

OVERALL LENGTH 91.7 FT.	A-THRUSTERS	E-EMERGENCY SPHERE	MAX. OPERATIONAL DIVING DEPTH 2000 F
OVERALL WIDTH 22 FT.	B-UNDERWATER HOUSE	F-CONTROL ROOM	MAX. UNDERWATER SPEED 4 KNOTS
OVERALL HEIGHT 28 FT.	C-LOCK CHAMBER	G-DIRECT SIGHT PILOTING	MAX. SURFACE SPEED 7 KNOTS
OP. DIVING DEPTH 1000 FT.	D-MACHINERY ROOM	H-LABORATORY	MAX. RANGE – SURFACE 400 NAUT. MILES

A Water Spider Made of Steel

Properly trained men supported by the appropriate equipment and technology can perform underwater almost any task they have been familarized with at the surface. This is true within limits that physiological research is extending: 1000 feet is already an operational depth, but we know that as long as the human body will need to breathe gases, even the lightest conceivable mixture will be too heavy to ventilate man's lungs at depths greater than 2000 feet.

Any deep dive in the sea needs careful preparation in hydropneumatic pressure chambers on land and requires costly, complicated, and cumbersome support facilities.

Helium-oxygen mixtures, saturation diving, undersea habitats, electronic lungs, heated suits, voice unscramblers, medical monitoring . . . there is a long list of indispensable, energy-consuming, vulnerable devices.

In 1965, after Conshelf III, we drew the lessons from the success that had needed so much money, so much surface manpower and equipment. The philosophy of our diving team had been, since the 1930s, that lines to the surface were the greatest danger for divers. The so-called lifelines had become potentially even more deadly with the increase in diving depth and the new need for great amounts of energy below. It became obvious to me that all the support complex had to migrate from the surface down to the immediate vicinity of the diver's habitat. The support systems had to leave the remote, unstable, stormy surface and go to the peaceful quiet of the ocean's deep bottom. The physiologist, the medical doctor, the instrumentation specialists would

The pressure-resistant steel spheres (below) that will make up the hull of the submarine Argyronete are maneuvered into position alongside a pier. Funding ceased before Argyronete could be completed.

live close to the oceanaut's quarters, only separated from them by an air lock and by viewing ports; but there would be a basic difference: they would be maintained at atmospheric pressure and would breathe re-generated air as in a submarine, while the oceanauts would be subjected to a pressure equal to that of the surrounding water, as in a Conshelf habitat. A further step in such an elaborate imitation of the water spider would be to include both quarters in a mobile, versatile submarine.

Our studies, conducted by engineer Jean Mollard and supervised by naval engineer Pierre Willm, ended in the construction of the ultimate in deep-diving equipment, a submarine bearing the name of our water spider, the *Argyronete*. This absolute weapon for ocean reentry is designed to service in safety four oceanauts working up to eight days, several hours a day, in 2000 feet of depth. The 300-ton *Argyronete* has a "dry" crew of six, a surface speed of seven knots, an underwater speed of four knots, and a cruising range of 400 nautical miles. It eliminates the need for a tender vessel.

The government suspended construction when it was more than three-quarters built.

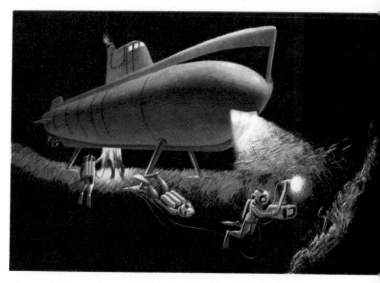

An artist's cutaway rendition of the **Argyronete** *(opposite), and another as it rests on the sea floor (above). Designed as a mobile Conshelf, it will vastly increase man's staying power on the bottom.*

Leaving aside the confusing lobbying and the political motivations, the real reasons for the suspension are that the oil companies (the only industry rich enough to fund such a project) were convinced that they could successfully achieve their deep jobs without divers, using remote-controlled tools. The concept, however, remains above criticism, and the *Argyronete* family of steel water spiders will multiply in the future.

Chapter IX. The Underwater Man

The speed and grace of a dolphin will never be matched by man for obvious anatomical reasons. Underwater, the dolphin has an additional advantage because its blood is able to store more oxygen than man's; it has body tissues that can withstand an oxygen debt; and carbon dioxide does not affect the dolphin's brain as severely as it does man's.

The human being has a better ability to remain underwater as soon as he is born than later in life. One wonders how a child would fare if a mother dared to give birth to her baby in water and rear him there.

Whatever ingenuity man uses to facilitate his reentry into the sea, his respiratory system provides him with few alternatives. He

---

"Due to the physical capabilities of his lungs, man is not able to breathe seawater directly."

---

can either enclose himself in a pressure-resistant, watertight shell in which he maintains his normal atmosphere or he can breathe more-or-less elaborate gas mixtures delivered to him at ambient pressures. In the first instance, he will have very poor contact with the ocean environment and, in the second, the physiological effects of breathing exotic gas mixtures under high pressure put a limit on his drive toward great depths. Training does not permanently improve man's ability to adapt to deep-sea diving; and centuries of a diving culture—such as exists among the oriental amas—has done nothing to alter the basic equipment genetically. The only breakthrough appears to be breathing without using gas at all. Experiments conducted on test animals have shown that liquid can be pumped through the lungs of mammals and sustain

life. But the liquid is a special concentration of salts with much, much more oxygen dissolved in it than normal ocean water. As we explain later, because of the incompatibilities in seawater and the physical capabilities of his lungs, there is no way that man can breathe seawater directly.

Hopes have been raised that perhaps man could "breathe" water indirectly, since small animals can live without difficulty for several days in a watertight enclosure made of Teflon membranes and completely submerged in an aquarium. Carbon dioxide accumulates inside the enclosure until its partial pressure becomes higher than that of carbon dioxide in the water. The gas then diffuses through the semipermeable membrane into the water and eventually into the air. In the same manner, as oxygen inside the enclosure diminishes, it is replaced by oxygen from the water. Thus the respiratory exchanges are assured. The atmosphere of the enclosure is just a little poorer than the air outside the tank. The very severe limitations of this solution will be reviewed and explained later. In the present stage of technology, no one sees how membrane regeneration could be adapted to great depths and to atmospheres of light gas mixtures.

In the future, patients suffering from lung cancer may be equipped with "artificial lungs," in which the blood would be regenerated through membranes by a special liquid. These patients then would become amphibians and would outperform their mammalian relatives who must still breathe air.

Water does not contain enough of essential dis-solved oxygen in order to sustain life in a warm-blooded animal such as man or dolphin.

Membrane Breathing

Respiration is always effected through membranes: the natural ones that constitute the walls of our lung alveoli and that separate air from blood; those of the capillary vessels that keep apart the blood from the tissues they irrigate; and in the case of fish, the membranes that coat the gills and separate blood from the seawater that bathes them.

Such live membranes are permeable to gases and are penetrated easily by their molecules, but contain liquids. Man today knows how to manufacture them; among artificial membranes are "silicone films" about five ten-thousandths of an inch thick, which were tested by W. L. Robb in 1964. These membranes are used specifically in such

*Silicone membranes, five ten-thousandths of an inch thick, make up the walls of this **hamster's cage.** Although watertight, gases move freely through the membranes in a direction governed by the partial pressures of the gas on either side of it. The one-atmosphere pressure inside the cage prohibits its use at common diving depths.*

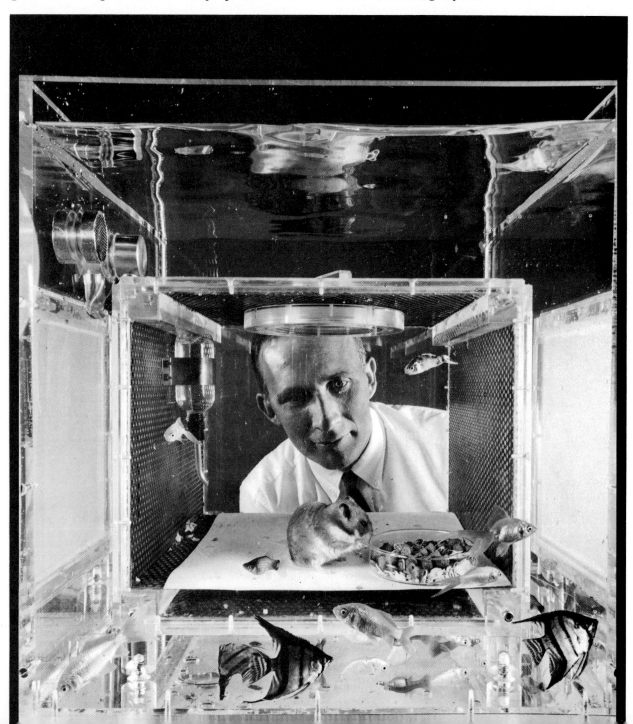

medical aids as artificial kidneys and it was quite a natural progression to next test them for underwater respiration.

If a tight enclosure made of this material contains a small animal such as a hamster, the animal will be able to breathe normally, as oxygen diffuses through the walls to replace the quantity that is used up, and carbon dioxide is eliminated in the same way. If such a "cage" is completely submerged in the water of an aquarium, it will carry on functioning normally. The watertight membrane keeps the enclosure dry and the exchanges of gases are maintained with the atmosphere, because gases also diffuse into the water. The surface of the sea or of a river thus functions as an immaterial membrane. This is moreover almost exactly what happens in an enclosed cave through which an underground river flows. Speleologists know very well that "wherever there is running water, there is breathable air; any excess of carbon dioxide is dissolved and eliminated."

For all these reasons, the partial pressure of each gas inside the "membrane cage" is identical to that of the same gases dissolved in the surrounding aquarium water and also the same as in the atmosphere. The cage can contain *only* air at atmospheric pressure. If any gases other than oxygen and nitrogen were introduced into it, they would escape to the atmosphere, and would be progressively replaced by air.

Pressure inside the enclosure being that of one atmosphere, the membrane walls must withstand the hydrostatic pressure of the surrounding water, an easy task in an aquarium, but positively impossible at any diving depth. Any conceivable wall construction material cannot at the same time have the required mechanical strength and porosity.

On the other hand, it is possible to regenerate an enclosure at atmospheric pressure

(such as a submarine, not a habitat) with seawater. The seawater is introduced in the enclosure, brought down to atmospheric pressure and sprayed in a "gas exchanger"; the water used is then ejected by a pump. At least a hundred gallons of water are necessary to extract about one quart of oxygen by this method. Regeneration of air by water spray gas exchangers has been proposed for military submarines; it requires great amounts of energy and it introduces a high degree of humidity into the submarine. This method of regeneration has yet to be used.

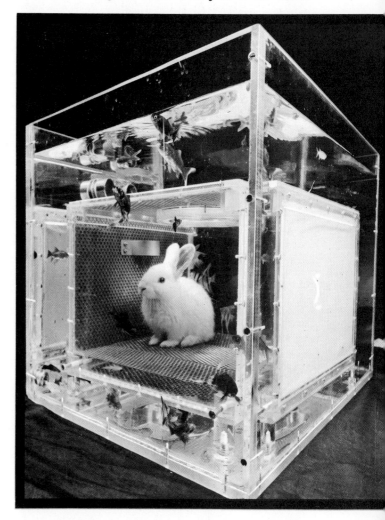

*The membranous walls of our lungs facilitate gas exchange between our blood and air; in other words, between a liquid and gas. In an identical manner, the **membranous walls of this rabbit's cage** allow for the exchange of oxygen and carbon dioxide between cage atmosphere and surrounding water.*

Breathing Water

Animal life arose in the sea and the oxygen needed to sustain it was drawn from the water. The simplest solution to man's problem of getting around in the ocean would seem to be having him breathe water like a fish. There is oxygen dissolved in the water, and the exhaled carbon dioxide could go into solution in return. Fish do this with their gills, highly vascularized structures supplied with blood rich in hemoglobin, a red pigment with an astonishing chemical affinity for oxygen. Fish simply exchange the oxygen they need and the carbon dioxide they expel directly between their blood and the water. In the human, however, the oxygen is taken up by the blood as it passes through the lungs at the same time carbon dioxide is deposited. The major difference, then, between the way a man and fish breathes is in the medium or carrier of the dissolved gases.

Fish are cold-blooded animals with very low metabolism; the oxygen requirements of warm-blooded mammals such as man are incomparably larger. To meet this need would require that enormous quantities of water be used, since seawater contains about one-thirtieth the amount of oxygen that air does. A man doing work needs about 21 quarts of air per minute to regenerate his blood with enough oxygen to meet his body's demand. To obtain the same amount of oxygen from ocean water as he does from those 21 quarts of air, a man would have to pump about 680 quarts of water a minute through his lungs. This would be a flow of 40 tons of water per hour—clearly impossible for our frail lungs.

There are other problems involved in breathing seawater. Oxygen, when dissolved in water, diffuses 6000 times more slowly than in air. The lungs, with their spherical alveoli, are able to exchange an easily diffusible gas, and are inferior to foliated gills for exchanges in liquids. Also, if the water temperature were below 98° F., there would be a problem of body heat loss. In addition, because the fluids in the human body are generally less saline than surrounding seawater, the resulting osmotic differences would cause the lungs to lose water and become dehydrated. For a number of reasons, then, man could not sustain himself by breathing seawater directly.

But Dr. Johannes A. Kylstra demonstrated that it was a simple matter to prepare a liquid that overcomes these difficulties. As early as 1961, at the University of Leiden, he performed some tantalizing experiments, in which mice and dogs have demonstrated the ability to breathe liquids that have salt contents similar to that of body fluids. In addition, these liquids had to be superoxygenated so that they would contain the same amount of oxygen as the same volume of air. The animals were pumping liquid through their lungs and lived. In some cases the animals were even able to return to breathing normal atmospheric air.

Some men being treated for lung ailments have had a lung irrigated with saline solution, and a volunteer diver "breathed" water in one lung while under anesthesia. It is theoretically possible, then, that both lungs could function like gills in special fluids and under special circumstances. If all problems could be solved, liquid-breathing divers of the future could penetrate much deeper than when breathing gases and rise to the surface rapidly without fear of gas bubbles forming in their bloodstream. Decompression illnesses and the need for expensive decompression chambers would be a thing of the past.

*Under certain laboratory conditions, **warm-blooded mammals can breathe liquids** (opposite). The solutions must be supersaturated with oxygen and have a salinity similar to body fluids.*

The Cancer Cartridge

Mammal lungs can function while "breathing" a liquid rich in oxygen, but the biggest problem is dealing with a buildup of carbon dioxide. Recently, additives absorbing carbon dioxide have improved such liquids, and it becomes theoretically possible to keep it at a near-zero concentration.

Anyway, in order to allow a person to function while "breathing" a liquid, a number of complex and delicate changes would have to be made. It would be necessary to surgically insert semipermanent injection tubes in the windpipe; provide heated diving suits; fill the lungs with an oxygenated, carbon-dioxide-absorbing liquid while the diver is under anesthesia; connect auxiliary tanks and circulatory pumps to the injection tubes in the windpipe; fill internal cavities with a saline solution; and maintain a standby team of medical assistants. Then such a person would be capable of diving for as long as six hours and as deep as at least 3000 feet. The diver would be permitted unlimited ascents and descents without any decompression, since no neutral gases would be breathed. At the end of the dive, the human would be anesthetized, the lungs emptied and revived with artificial respiration. After the equipment is removed, further care would have to be given in order to avoid pulmonary infection so that the lungs, which have been deprived of indispensable surfactants in the flushing, may return to normal.

An even more daring solution, though, than fluid breathing is a blood exchanger, a cartridge which would allow direct exchange with the blood. Artificial kidneys and mechanical heart-lungs are already part of medical science history. The current equipment is large and bulky. When it can be miniaturized, it would constitute an artificial gill made of incompressible solids and fluids.

Being able to function independently of pressure, it could establish a favorable exchange between blood and a special solution with the ability to oxygenate as well as absorb carbon dioxide. Such a chemical solution would be automatically maintained to optimum efficiency by electronic controls from oxygen and chemical reserves.

A person equipped with this system is very close to being *Homo aquaticus*. Corrective lenses would be added for improved water vision and air cavities could be filled with any harmless fluid.

Once medical science perfects equipment for lung cancer patients, it could be used by normal persons equipped with only arteriovenous offtakes, like those worn by artificial-kidney users. The lungs would be only partially filled with fluid, enough to avoid collapsing of the chest, and breathing reflexes would be inhibited by drugs. The underwater performances of men so equipped would be superior in depth and duration to those of the best-adapted cetaceans.

The lung substitute will enable divers to descend to any depth and return without decompression. Blood will be pumped through a highly oxygenated medium; the exchange of gases will take place through a semipermeable membrane. A computerized motor activator will respond to sensors detecting the gases dissolved in the blood and adjust the flow of oxygen (O_2) and the absorption of carbon dioxide (CO_2). A recharged cartridge can easily be substituted periodically.

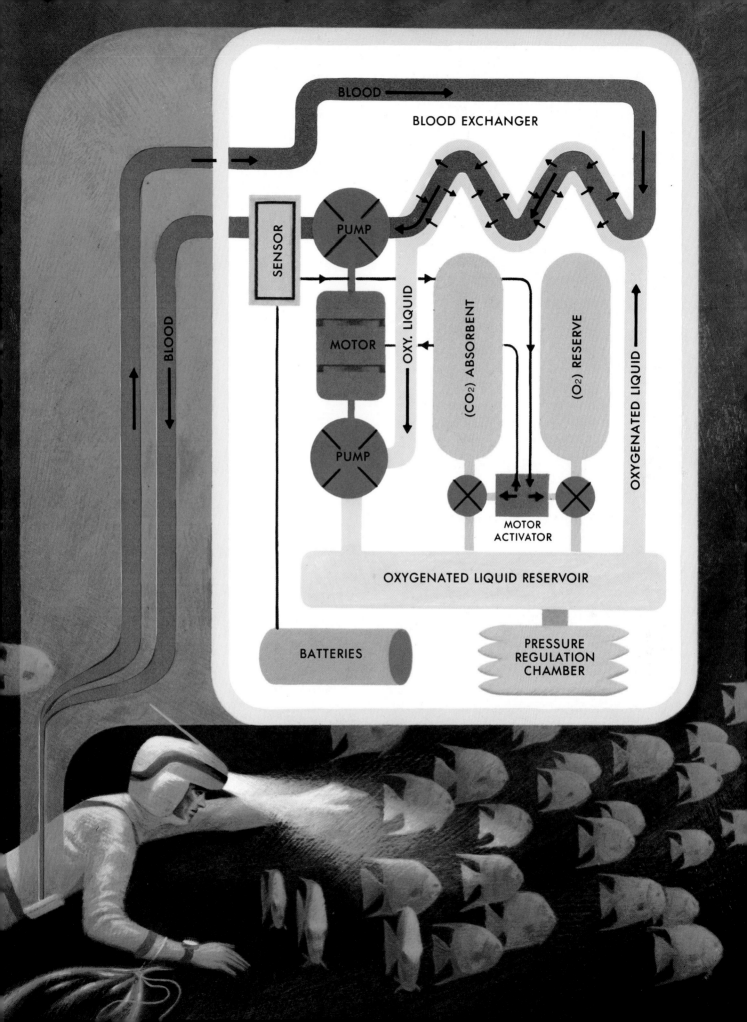

Invasion of the Sea?

Man developed terrestrial living to its highest form, but perhaps he succeeded too well, for now he has produced more people than the earth can support and has developed the means to destroy himself, if not the whole world. Today, he is perfecting methods and equipment to perform a spectacular reentry into the sea.

Man would certainly not be the first land animal to return to mother sea. Long ago, in

The sea has been home for animals (above) as long as life has existed. Others, like cetaceans, have returned. Is man beginning his reentry (right)?

the geologic past, reptiles made their home in the sea after a time on land—these were creatures such as the fishlike ichthyosaurs and plesiosaurs. Even today we can see turtles, sea snakes, and the remarkable marine iguanas of the Galápagos making their accommodation with the ocean. Being a warm-blooded mammal should not deter man, for his cousins the whales and dolphins wander about the globe totally at home in the water world. Seals and walruses represent an intermediate step, spending part of their lives on land and certainly not feeling out of their element in the sea.

From the naked divers of 5000 years ago up to the present, man's progress undersea has been very slow and has suddenly accelerated with the birth of technology. His adaptation is thus artificial; it is not the result of anatomical and physiological evolution, but it is one of the prints of the development of his mind. Sports and games, science and engineering have outlined routes to take. Now the pragmatists take over. The petroleum industry has been the first to really exploit the world under the sea. Those soon to follow include mineral-mining interests. Sea farms, cultivated from beneath the surface, can't be far behind since the need for food is a major driving force. Hard hats and aqualungs have made short incursions possible, followed by exploration subs; but the actual occupation of the ocean bottom has only tentatively begun with the advent of underwater habitats and lock-in-lock-out submarines.

As early as the year 1930, Sir Robert Davis envisioned men living on the bottom of the sea at atmospheric pressure, with an air-lock entrance as an access to the ocean. Oil companies of 1974 prepare such atmospheric-breathing, pressure-resisting chambers to be clamped on drilling machinery working on the deep ocean floors. Other intruders in the ocean, directly a product of man, belong to the family of remote-controlled robots. They are no match for a diver, when depth does not exceed our physiological limits; but they outperform humans in depth and duration. Robots and divers will coexist in the future. Both contribute, directly or indirectly, to the human invasion of the ocean and our quest for knowledge.

Whether human intrusion in the fragile sea is advisable will be discussed in another book, but it is happening under our eyes. Science fiction writers dream of entire populatons moving permanently to underwater cities, developing diverging civilizations and making war against their land brothers. No. Settlements will be built on the sea floor and inhabited as long as a definite task justifies the cost and the sacrifices. But *Homo aquaticus* himself, after a few weeks of work in platinum mines 7000 feet deep, will gladly rejoin his native village, celebrate a friend's birthday, and smile in the warm sun.

Drawing by Ed Fisher; Copr. 1955, The New Yorker Magazine, Inc.

Messenger of the Sun

Born head on to my weight, Messenger of the Sun,
I dive, a torch in hand, to the heart of the sea.
Bringing a load of air, eager, without a gun,
Through twilight provinces begins my odyssey.
Shark, tuna, and dolphin—in a hunting garden—
Are still creatures of light soon lost in oblivion.
I think deeper—below

Incessant clouds of snow
Embed wrecks of galleons
In silent orisons.
A dusty vault, realm of clingers, crawlers, diggers,
Tunneling through eons of graves. And scavengers
Expecting manna of death to rain from heaven,
Transmitting death into far promises of life.

Miles down, the gospel from days, seasons, moon and sun
Is told in the lands where eternal dark is rife
By chains of tiny messengers.
The hunger for fresh food triggers

A series of short trips, upward back to the sun.
Into the dun vastness, where sparks and stars are none,
Where jagged rocks emerge from plains of skeletons,
Messenger of the Sun, the human mind has gone.

Index

ILLUSTRATIONS AND CHARTS:

Howard Koslow—32-33 (bottom), 53, 79, 131 (top), 138-139.

PHOTO CREDITS:

The Bettmann Archive, Inc.—71; H. Broussard—54; O. Buselli (Courtesy of Technisub—56; Culver Pictures, Inc.—64; Free-lance Photographer's Guild: Dennis Hallinan—15 (bottom), Scott Penwarden—15 (top); General Electric Re-entry and Environmental Systems Division—134, 135; Claude Houlbreque—88 (right); Imperial War Museum, London—32 (top), 74; Mariners Museum, Newport News, Va.—39 (bottom); Giulio E. Melegari—57; Musée de la Marine, Paris—38 (bottom), 84 (left); NASA—16 (top); National Oceanic Atmospheric Administration, Manned Undersea and Technology—105 (bottom, left), 122, 128, 129; Naval Photographic Center—116, 117 (middle); W. J. O'Neill, Dive Director, Westinghouse Electric Corporation—105 (top), 124; Edward S. Ross—60; The Sea Library: B. Campoli—48 (bottom), 67 (top), 69 (top), 114, 115, T. J. Doty—17 (left), William L. High—19, Jack McKenney—5, 6, 17 (right), 22, 23, 86, Robert Marx—16 (bottom), Carl Roessler—27, 91, Joseph A. Thompson—41, 47 (top), 59, 126 (top), Paul Tzimoulis—90; Siebe, Gorman & Company Ltd. (from Deep Diving and Submarine Operations by Sir Robert H. Davis)—36-37, 39 (top), 40, 61, 65, 67 (bottom), 68, 76 (top); Tom Stack & Associates: Ron Church—11, 47 (bottom), 117 (top), 125 (bottom), Ben Cropp—80, Al Giddings—104, 105 (bottom, right), 123 (top), 126 (bottom, right); Taurus Photos: Dick Clarke—18, 83, Bob Dunn—25, 52, Dave Woodward—14, 72-73 (bottom), 92 (right), 123 (bottom), 136; Paul Tzimoulis—102, 103; U.S. Navy—49, 117 (bottom).